HOMESTEADING ESSENTIALS

A Practical Guide to Growing Food, Raising Animals, and Achieving Sustainable Self-sufficiency for Everyday Living

LANDON WOODLAND

Copyright © 2024 by Landon Woodland

All rights reserved.

No part of this book may be reproduced in any form without the written permission of the publisher or author except as permitted by U.S. copyright law.

CONTENTS

1 WELCOME TO THE HOMESTEADING LIFE
    1 My Story: From Suburbia to Homestead Living
    2 Reconnecting with Nature and Achieving Self-Sufficiency
    2 What This Book Will Teach You
    3 Start Small, Grow Big

5 WHERE TO BEGIN
    5 Understanding Your Space: Urban, Suburban, and Rural Homesteading
    7 Essential Tools and Resources to Start
    12 Setting Realistic Goals for Your First Year
    13 Avoiding Common Beginners Pitfalls

16 GROWING YOUR OWN FOOD IN SMALL SPACES
    16 Container Gardening: Vegetables, Herbs, and Fruit
    19 Vertical Gardening: Maximizing Space Efficiently
    23 Raised Beds vs Traditional Gardeninig: Which is Right for You?
    27 Hydroponics for Small Spaces
    29 Bringing It All Together: Choosing Your Gardening Method
    30 Gardening on a Budget: DIY Tools and Affordable Seeds

CONTENTS

34 RAISING SMALL LIVESTOCK FOR BEGINNERS
34 Best Livestock for Small Homesteads: Chicken, Rabbits, Bees
39 Legal Considerations and Space Requirements
42 Basic Care: Feeding, Housing, and Maintenance
45 Simple, Low-Cost Livestock Housing Solutions

48 SUSTAINABILITY ON A BUDGET
48 Planning a Low-Cost Homestead: Saving Without Sacrificing Quality
50 DIY Projects: Building a Compost Bin, Rainwater Collection Systems, and More
59 Thrift Store Hacks for Homesteaders
61 Creative Ways to Save Energy and Resources

65 THE ART OF FOOD PRESERVATION
65 Canning: Your Step-by-Step Guide
68 Freezing, Dehydrating, and Pickling Produce
71 How to Store Your Preserved Foods Safely
74 Creative Uses for Preserved Goods in Everyday Meals

78 HOMESTEADING FOR BUSY PEOPLE
78 Time-Saving Tips for Weekends and Busy Schedules
80 Efficient Gardening Methods (No-Dig and Low-Maintenance Approaches)
83 Quick Livestock Care Routines
85 Staying Organized: Daily, Weekly, and Seasonal Checklists

89 SEASONAL HOMESTEADING: WHAT TO DO EACH SEASON
89 Spring: Preparing for Growth
92 Summer: Nurturing and Managing
95 Fall: Harvest and Preservation
98 Winter: Maintenance and Planning

103 EMERGENCY PREPAREDNESS AND HOMESTEAD SECURITY

103 Stockpiling Essentials for Emergencies
104 Protecting Your Homestead: Basic Security Measures
105 Power Backup Solutions
106 Preparing for Natural Disasters and Extreme Weather

108 GROWING YOUR HOMESTEADING SKILLS

108 Expanding Your Garden and Livestock: Intermediate Skills
110 Learning New Crafts: Soapmaking, Herbal Remedies, and Sustainable Cleaning Products
114 How to Start Generating Income from Your Homestead
116 Experimenting with New Homesteading Techniques

118 EMBRACING YOUR ADVENTURE AHEAD

118 The Power of Connection
119 You're on the Right Path—Keep Going!

121 ACKNOWLEDGMENTS

WELCOME TO THE HOMESTEADING LIFE

My Story: From Suburbia to Homestead Living

Welcome, my name is Landon Woodland. I never imagined trading my suburban life for a fulfilling homesteading adventure. Like many, I was used to the conveniences of city living—grocery stores around every corner, a manicured lawn that required minimal attention, and the busy rhythm of daily life. Yet, something was always missing. I longed for a deeper connection with nature and a sense of accomplishment from working with my hands and providing for myself. That's when I began exploring the idea of homesteading.

My journey started small: a few herb pots on the windowsill, then a modest vegetable garden in the backyard. The more I learned about growing my own food, the more excited I became. What began as a hobby quickly turned into a passion. I was hooked. Eventually, I traded my city job and suburban comforts for a life centered around self-sufficiency. It wasn't always easy, but every challenge taught me something valuable.

This change gave me a new perspective on life that focuses on sustainability, hard work, and the joy of living simply. It taught me that anyone—even on a tiny urban balcony—can embrace the homesteading lifestyle. In this book, I'll guide you through the process, sharing practical steps to help you start your own homestead, no matter where you are today.

Reconnecting with Nature and Achieving Self-Sufficiency

Homesteading offers a unique chance to reconnect with nature and embrace a simpler, more intentional lifestyle. In today's world, where many of us are far removed from our food sources and the natural world's rhythms, it brings us back to the basics. It allows you to witness firsthand the cycle of life—from planting seeds to harvesting crops, from caring for animals to enjoying the fruits of your labor.

But the beauty of this lifestyle goes beyond just growing your own food. It's about achieving a level of self-sufficiency that brings both freedom and peace of mind. When you learn how to produce, preserve, and sustain what you need, you rely less on outside systems and feel more empowered in your everyday life. This shift doesn't have to happen overnight. It's a progress that gradually allows you to build skills and knowledge over time. You'll reconnect with the world around you and your true self. Over time, you'll find that you have achieved a deeper understanding of your capabilities and resilience.

What This Book Will Teach You

This manual is designed to be your **practical guide** to starting and thriving in homesteading, no matter where you begin. Whether you live in the city, the suburbs, or a more rural setting, the principles and steps in this book will apply to your situation. I've broken everything down into easy-to-follow sections so the process won't overwhelm you.

Each chapter will walk you through **key aspects**, from growing your own food to raising small animals preserving what you produce, and preparing for potential challenges like emergencies or extreme weather. The goal is to ensure you feel empowered to take small steps first and then gradually build your skills and confidence.

We'll start with understanding your space and how to make the most of it, even if you're working with a small balcony or backyard. Then, we'll dive into the tools you'll need and the essential tasks to focus on in your first year. I'll also guide you through the more hands-on aspects, like learning how to care for animals and preserve your food and exploring ways to make your homestead more sustainable without breaking the bank.

Along the way, you'll also learn how to fit homesteading into a busy schedule, allowing you to evolve at your own pace without sacrificing other aspects of your life. The focus will be on **actionable steps**, supported by real-world tips, so you can immediately implement what you've learned.

By the end of this reading, you'll have a solid foundation in homesteading, armed with practical knowledge and a sense of confidence to continue expanding your homestead over time.

Start Small, Grow Big

Starting your journey might initially feel overwhelming, but here's a key principle that will keep you grounded: **start small and grow big**. Homesteading isn't about doing everything at once. It's about taking small, meaningful steps that gradually build your confidence and skills over time. The best part? Each small success will motivate you to keep going.

It's important to remember that this is not a race. There's no need to rush or compare yourself to others who may already have thriving gardens or livestock. Whether you're growing herbs on a windowsill or tending to a small garden in your backyard, every bit of progress counts. In fact, many experienced homesteaders will tell you that starting small is the smartest way to begin, as it allows you to learn and adapt without getting anxious and feeling like you can't do it.

The secret to success is **patience and persistence**. You'll have good days, and you'll have challenging ones. But that's part of the learning process; each obstacle is an opportunity to gain valuable experience. Each season, you'll see improvements in your skills, and what once seemed daunting will start feeling second nature. Now, it's time to take the first steps. Don't worry about having all the answers right away—I'll guide you through each phase from the very beginning.

Embrace the journey, stay curious, and you'll grow more than you ever thought possible before you know it. The adventure ahead is exciting, and you're ready to begin!

2
WHERE TO BEGIN

Let's dive into the practical side. Every successful homestead starts with a solid plan: the key is to evaluate your space and find creative ways to maximize its potential. Whether you're working with a small urban garden, a suburban backyard, or acres of rural land, homesteading can be tailored to fit your environment. By the end of this chapter, you'll have a clear plan that matches your unique situation and aspirations.

Understanding Your Space: Urban, Suburban, and Rural Homesteading

Urban homesteading often involves limited outdoor space, such as a balcony, rooftop, or even just a few sunny windows. Don't let that discourage you—urban homesteaders can still grow a surprising amount of food and make meaningful steps toward self-sufficiency. *Container gardening* is an excellent solution for those with restricted space. You can grow herbs, small vegetables, and even dwarf fruit trees in pots or raised containers on balconies or rooftops. *Vertical gardening* is another powerful technique, allowing you to grow plants upward using trellises, hanging baskets, or wall-mounted planters. *Hydroponics* is another space-saving technique that allows you to grow plants in nutrient-rich water, perfect for urban settings where outdoor soil may be limited.

In addition to gardening, urban homesteaders can take advantage of *indoor projects* like growing microgreens, cultivating herbs, or even keeping bees if local regulations allow it. You can also focus on creating a more sustainable household by making homemade cleaning products, conserving water, or composting food scraps, even on a small scale. We'll explore these techniques, as well as strategies for suburban and rural homesteading, in later chapters, providing you with the practical know-how to succeed in various environments.

Suburban homesteading offers more space than urban environments but usually comes with limitations compared to rural living. If you have a backyard, you can start with a small garden bed or raised beds, depending on the size of your yard. Suburban homesteads are ideal for those wanting to experiment with both *food production and raising small livestock*. With a bit of research into local zoning laws, you may be able to keep chickens, rabbits, or bees in your backyard.

Maximizing space in a suburban setting requires smart planning. You might combine *edible landscaping* with traditional landscaping, integrating fruit trees or berry bushes into your garden design. *Water conservation* is another important consideration for suburban homesteads. Installing a rain barrel or using drip irrigation systems can help reduce water waste while keeping your garden thriving. We'll delve into these strategies later in the book, offering step-by-step instructions to get you started.

Rural homesteading usually offers more *land to work with* but also requires more maintenance and planning. With larger areas, you have more options for growing food, keeping livestock, and even implementing sustainable energy systems such as solar panels or wind turbines. However, rural homesteads often require a deeper understanding of your land's specific characteristics—such as soil quality, water sources, and climate conditions.

If you're working with several acres, you can plan for larger gardens, greenhouses, or even small orchards. *Livestock options* also expand in rural settings. In addition to chickens, you may choose to raise goats, sheep, or pigs, depending on your goals and experience. Rural homesteaders also have the advantage of space for *larger-scale projects* like building barns, creating ponds, or dedicating areas for permaculture.

Despite the space, rural homesteaders must still consider the efficient use of resources. *Water management* is crucial, especially if you rely on well water or rainwater harvesting. Ensuring you have access to reliable water sources for both your household and animals is essential for long-term sustainability. We'll cover these large-scale projects in more detail as we progress through these pages.

Essential Tools and Resources to Start

Once you clearly understand your space, the next step is gathering the fundamental equipment needed to start your homesteading journey. Having the right asset from the beginning can save you time, energy, and frustration. In this section, we'll explore the must-have tools that will help you tackle your first projects and make your day-to-day tasks more efficient.

Gardening Tools

Whether you're working with containers on a balcony or a backyard garden, a good set of gardening tools is a must. Here are a few essentials to get started:

- **Hand trowel**: Ideal for planting small plants, herbs, or bulbs, and it's versatile enough to be used in small garden beds or containers.
- **Pruners**: A quality pair of pruners will help you trim and maintain plants, ensuring they stay healthy and continue to produce.
- **Watering can or hose**: Depending on the size of your space, choose between a watering can for smaller gardens or a hose with a spray nozzle for larger areas. Efficient watering is key to keeping your plants hydrated without wasting water.
- **Gloves**: A durable pair of gardening gloves will shield your hands from thorns, soil, and any potential hazards, such as blisters from handling tools.

As you expand, you might consider adding a **garden fork**, **hoe**, and **rake** for larger plots, but these basics are more than enough to get started.

Tools for Raising Livestock

If you're planning to raise small livestock like chickens or rabbits, there are a few essential tools you'll need to make daily tasks easier:

- **Feeders and waterers**: Ensure that your animals have constant access to fresh food and water. There are various styles available, from gravity feeders to automatic waterers, so choose one that fits your setup.
- **Cleaning tools**: Keeping your animal enclosures clean is critical for their health. A sturdy broom, rake, and shovel will help you maintain cleanliness with ease.
- **Basic fencing tools**: Even for small animals, good fencing is essential. To build or repair enclosures, you may need tools like wire cutters, pliers, and zip ties.

DIY and Maintenance Tools

Homesteading often involves a lot of DIY projects, particularly when it comes to building or maintaining structures like raised garden beds, compost bins, or animal shelters. Here's a basic toolkit you'll want on hand:

- **Hammer and nails**: Whether you're fixing a fence or building a simple structure, these are essential.
- **Drill and screws**: A cordless drill can save you time and effort on various projects. Having a set of screws handy will make construction and repairs smoother.
- **Measuring tape**: Precision is key when building enclosures, garden beds, or any homesteading structures, so keep a tape measure close by.
- **Saws**: Depending on your level of comfort, a hand saw or circular saw can help you cut wood for DIY projects.

Sustainable Resources

Homesteading is also about sustainability, so gathering the right resources from the start can set you on the right path:

- **Compost bin**: Whether purchased or homemade, composting allows you to reuse kitchen scraps and yard waste, providing nutrient-rich compost for your garden.
- **Rainwater harvesting system**: Collecting rainwater in a simple barrel or more advanced system can help reduce your reliance on municipal water sources.

- **Seed library or heirloom seeds**: Starting with high-quality, non-GMO seeds is a great way to ensure your plants will grow healthy and strong.

Testing Your Soil

Before you begin planting, it's essential to understand the condition of your soil. Soil testing allows you to identify the pH level and nutrient content, giving you valuable insights into whether your garden will thrive as is or if amendments are required to improve fertility. Knowing your soil's health can prevent problems like nutrient deficiencies or poor plant growth, making this an important first step for any homesteader.

Why Testing Your Soil Matters

The results of a soil test help you understand if your soil has the right balance of nutrients—such as *nitrogen* (N), *phosphorus* (P), and *potassium* (K)—and whether the pH level is suitable for the crops you want to grow. Different plants thrive in different conditions, and correcting issues with pH or nutrient levels before planting will save you time and effort down the line.

How to Test Your Soil at Home

Testing your soil can be done easily with **store-bought soil test kits** or by sending a sample to your local agricultural extension for a more comprehensive analysis. Here's how you can do it yourself:

- **Gather Soil Samples**: Take samples from several spots in your garden to get an accurate overall reading. Dig down about 6 inches for each sample, and collect soil in a clean container. Mix the samples together to form one composite sample.
- **Test for pH and Nutrients**: Use a *soil test kit* to evaluate the pH level and nutrient content. The kit typically comes with simple instructions—add the soil to test tubes, mix with water or a provided solution, and compare the color result to the pH or nutrient chart.

Understanding Your Results:

pH: Most vegetables and fruits prefer a mildly acidic to neutral pH (between **6.0 and 7.0**). Adding **lime** can raise the pH if your soil is too acidic, while **sulfur** can lower it if it's too alkaline.

Nutrient Levels: Your test will indicate whether your soil has enough nitrogen, phosphorus, and potassium. If the levels are low, you can improve your soil by adding organic materials such as compost, manure, or targeted fertilizers.

Advanced Testing and Professional Analysis

You can send your soil sample to a local agricultural extension or professional lab for a more in-depth understanding. They will provide a detailed breakdown of your soil's composition, including trace minerals, organic matter content, and more specific recommendations for amending your soil.

Testing your soil at the start ensures a strong foundation for your garden. Armed with this knowledge, you can make informed decisions about which amendments to use, helping your plants thrive and grow more effectively.

Setting Realistic Goals for Your First Year

When you're at the beginning, it's easy to feel excited and want to tackle everything at once. But the key to long-term success is to *set realistic goals*, especially in your first year. By breaking down your ambitions into manageable tasks, you'll avoid burnout and give yourself the best chance of achieving your goals. Remember, homesteading is a gradual process, and the journey is just as important as the destination.

The first step in setting realistic objectives is to **evaluate your priorities**. Ask yourself: *What's most important to you? Is it growing your own food? Raising livestock? Becoming more self-sufficient in energy and water use?* You can focus your efforts on what matters most by identifying your top priorities. For example, if your main goal is to grow your own vegetables, then your first-year plan might concentrate on creating a small garden and learning about soil health, planting, and harvesting.

It's also important to consider your **available time and resources**. Homesteading can be time-consuming, so be realistic about how much time you can dedicate each week. If you're balancing homesteading with a full-time job, for example, it's better to start small and expand as you get more comfortable with the daily tasks.

In your first year, it's best to focus on **one or two core projects** rather than trying to do everything at once. For example, you might start with a small garden and gradually add more plants as you gain experience. If you're considering raising animals, you might begin with a small group of chickens or rabbits. By focusing on just a few projects, you'll be able to dedicate the time and attention needed to learn and grow without feeling overwhelmed.

As you select your projects, keep in mind that **learning through doing** is a core principle of homesteading. You're bound to make mistakes along the way, but that's part of the process. Embrace the learning curve, and don't expect perfection in your first year. With each season, you'll gain more knowledge and confidence.

Break down your yearly goals into **seasonal milestones** to stay organized and on track. For example, if your goal is to grow your own vegetables, your spring milestone might be preparing the garden beds and planting seeds, while your summer milestone could be caring for the plants and harvesting early crops. Breaking your goals into smaller, seasonal tasks makes them more manageable and helps you stay motivated as you see progress throughout the year. We'll delve into this important concept in a dedicated chapter.

Keep a journal or planner to track your milestones and progress. This will help you stay focused and give you valuable insights for future seasons. If something didn't go as planned, you can learn from it and make adjustments next year.

As you move through your first year, it's important to **celebrate small wins**. Whether it's harvesting your first batch of vegetables, successfully building a compost bin, or watching your chickens lay their first eggs, these achievements are significant. Acknowledging your progress will keep you motivated and remind you why you started this journey in the first place.

Focusing on these practices will make your first year of homesteading productive and rewarding. Now, let's talk about some common beginner pitfalls and how to avoid them so you can keep moving forward confidently.

Avoiding Common Beginner Pitfalls

As exciting as starting your homesteading journey can be, it's easy to fall into a few common challenges along the way, especially as a beginner. The good news is that you can avoid these setbacks with the right approach and confidently move forward. Learning from others who have been through the same obstacles can prevent frustration and make your first year a smoother, more rewarding experience.

. . .

1. Starting Too Big, Too Fast

One of the most common mistakes beginners make is trying to do too much all at once. The enthusiasm to grow your own food, raise animals, and become completely self-sufficient is understandable, but tackling too many projects at once often leads to burnout. As we discussed earlier, starting small and gradually expanding is crucial as you gain confidence and experience. Starting with one or two manageable projects allows you to focus your time and energy, ensuring each task is done properly. Master one area before moving on to the next.

2. Underestimating Time and Effort

Homesteading requires time, patience, and effort. Many beginners underestimate the amount of time needed to maintain a garden, care for animals, or manage homestead projects. It's important to be realistic about how much time you can commit each day or week, especially if you're balancing homesteading with a full-time job or family responsibilities. Setting realistic time expectations ensures that self-sufficiency doesn't become daunting and allows you to enjoy the process.

3. Ignoring Local Laws and Regulations

It's essential to check local laws and regulations before starting certain homesteading projects, particularly if you plan to raise animals. Some areas have restrictions on keeping livestock like chickens, rabbits, or bees, especially in suburban or urban neighborhoods. Avoid the frustration of having to dismantle your hard work by doing thorough research before starting. Reach out to your local government or community groups for advice on what's allowed in your area and any permits you might need.

4. Neglecting Soil Health and Water Management

Another common pitfall is not paying enough attention to soil health and water management. Beginners often assume they can plant in any patch of soil, only to find that their plants struggle or fail. Healthy, nutrient-rich soil is the heart of a successful garden, so take the time to test your soil and enrich it with compost or organic matter if needed. Similarly, ensure you have a water plan in place. Whether you're using a hose, drip irrigation, or rainwater collection, consistent watering is essential for your plants to thrive. Neglecting these basics can result in wasted time and effort, so it's worth investing time early on to understand your soil and water needs.

5. Expecting Immediate Results

Homesteading is a process, and it takes time to see the fruits of your labor—literally. Many beginners feel discouraged when they don't get immediate results or when things don't go as planned. It's important to remember that setbacks are part of the learning experience. Plants may not grow as expected, or livestock may require more attention than anticipated. The key is to stay patient and embrace the learning curve. With each season, you'll see improvement, and the small victories will add up over time.

By being mindful of these common traps and taking a patient, measured approach, you can set yourself up for a successful first year in homesteading. In the next chapter, we'll dive into the rewarding experience of growing your own food, with practical advice on how to start small and maximize your space.

3
GROWING YOUR OWN FOOD IN SMALL SPACES

Container Gardening: Vegetables, Herbs, and Fruit

Container gardening is one of the most practical and versatile ways to grow your own food, especially if you're working with limited space. Whether you're living in an urban apartment with a balcony or a suburban home with a small yard, it allows you to grow fresh vegetables, herbs, and even fruit in manageable, moveable pots and planters. Best of all, it's an excellent way to get started, as it requires fewer resources than traditional gardening and allows you to control soil quality, sunlight, and water more easily.

Why Choose Container Gardening?

Container gardening is particularly suited for those with limited space, but even those with larger areas can benefit from this method. One of the main advantages is *flexibility*—if a plant isn't getting enough sunlight in one spot, you can easily move it to a sunnier location.

Containers also allow you to control soil conditions more precisely, meaning you can grow plants that might not thrive in your native soil. Additionally, these gardens are easy to maintain and simplify managing pests and diseases that could otherwise spread throughout a larger garden bed.

Container gardening offers a great entry point because it reduces some of the common issues that new gardeners face, such as poor soil or drainage problems. You can start small, with just a few pots, and expand as you gain confidence.

Choosing the Right Containers

The first step in successful container gardening is choosing the right containers for your plants. Here are a few factors to consider:

- **Size**: The dimensions of the container will depend on what you're planning to grow. For example, herbs and small plants like lettuce can thrive in smaller pots, while larger plants like tomatoes or peppers need deeper and wider containers to accommodate their roots.
- **Material**: Containers can be made from different materials, including plastic, terracotta, ceramic, or even upcycled materials like buckets or old tubs. Each substance has its pros and cons. Plastic pots are lightweight and easy to move but may not provide the best insulation for the soil. Terracotta and ceramic pots offer better insulation but can be heavier and prone to breaking.
- **Drainage**: It's essential to ensure your containers allow excess water to escape. Water can pool at the bottom without proper flow, causing root rot and other issues. If your containers don't have drainage holes, you can add them using a drill or by placing a layer of rocks at the bottom to allow water to flow away.

Selecting the Right Plants

When choosing plants for your container garden, it's important to select varieties that are well-suited for container growing. Many vegetables, herbs, and fruit varieties are specifically bred for small spaces. Here are some great options to start with:

- **Herbs**: Aromatic plants like basil, thyme, rosemary, mint, and parsley are ideal for container gardening. They don't require much space, and many herbs can thrive with moderate sunlight and regular watering. You can keep them on a windowsill or in a small pot on your balcony.
- **Vegetables**: Leafy greens like lettuce, spinach, and kale are excellent choices for beginners. They grow quickly and don't require deep containers. Tomatoes, peppers, and cucumbers can also thrive in containers but will need larger pots and stakes or cages for support.
- **Fruit**: Dwarf fruit trees, like lemon or fig trees, are optimal for container gardening and can be grown on a balcony or patio. Strawberries are another popular option, as they can be grown in hanging baskets or shallow containers.

Soil and Fertilizer

Container plants depend heavily on the quality of the soil you provide, so it's important to choose a good **potting mix**. Unlike garden soil, the potting mix is lighter and provides better drainage. You can also find mixes enriched with compost or organic matter to provide essential nutrients.

Since the nutrients in containers can get depleted more quickly than in garden beds, you'll need to supplement your plants with **fertilizer**. Organic ones, such as compost tea or fish emulsion, work well for container plants and help maintain soil health over time. Be sure to follow the instructions on the fertilizer label to avoid over-fertilizing, which can damage your plants.

Watering and Sunlight

Container plants tend to dry out more quickly than those in garden beds, so it's important to water them regularly. The key is to keep the soil consistently moist but not waterlogged. Check the soil daily by sticking your finger an inch or two into the soil—if it feels dry, it's time to water. During the hotter months, you may need to water your containers more frequently, especially for plants in direct sunlight.

Regarding sunlight, most vegetables and herbs need at least 6 hours of direct sunlight daily. If you're working with a balcony or patio, take note of how the sun moves throughout the day and place your containers in the best spot to receive adequate light. For areas with less sunlight, you can opt for shade-tolerant plants like spinach, lettuce, or certain herbs like mint and parsley.

Vertical Gardening: Maximizing Space Efficiently

For those with limited ground space, vertical gardening is a highly effective way to grow more plants by utilizing upward space. This technique is particularly useful for urban or suburban homesteaders with small yards, balconies, or even patios. It involves growing plants on trellises, walls, or stacked containers, making it an ideal solution for small spaces while adding an attractive, green element to your home. Whether you're cultivating vegetables, herbs, or flowers, vertical gardening offers a practical and creative way to maximize your space efficiently.

Why Vertical Gardening?

The key advantage of vertical gardening is its ability to **maximize limited space**. Traditional gardening requires horizontal ground space, but by growing plants upward, you can increase the number of plants you grow without taking up more square footage. This method also improves air circulation, helps prevent some soil-borne diseases, and makes harvesting easier as plants are raised to a more accessible height. Additionally, it can serve a decorative purpose, transforming dull walls or fences into lush green spaces.

It requires relatively low maintenance once established. It allows you to grow various crops in a small footprint and can easily be adapted to your environment.

Types of Vertical Gardening Systems

There are several ways to implement vertical gardening, depending on your space, the plants you want to grow, and your budget. Here are a few popular systems to consider:

- **Trellises and Arbors**: A *trellis* is one of the most common methods for vertical gardening. It provides support for climbing plants like cucumbers, beans, peas, or squash, allowing them to grow vertically instead of spreading across the ground. You can build a simple trellis from wood or metal and place it in your garden beds, containers, or against a wall. *Arbors* are similar but often more decorative, creating a shaded walkway while supporting vines and other climbing plants.
- **Hanging Planters and Baskets**: Perfect for balconies or small patios, hanging baskets and planters allow you to grow

plants off the ground. Herbs like parsley, cilantro, and thyme do particularly well in hanging planters, as do flowers like petunias or nasturtiums. Hanging planters also add visual appeal and can be hung from walls, railings, or hooks.
- **Vertical Towers or Stacked Planters**: These systems consist of stackable containers or towers where multiple layers of plants can grow upward. Strawberries, lettuces, and herbs are well-suited to this style of gardening. Some vertical towers come with built-in watering systems that distribute water evenly to all levels, reducing the time you spend watering.
- **Wall-Mounted Gardens**: For extremely small spaces, wall-mounted gardens are ideal. You can create pockets for soil using fabric or containers mounted directly onto a wall, allowing you to grow herbs, flowers, or small vegetables. This is perfect for a sunny wall on a balcony or patio, providing a space-saving solution that turns otherwise unused vertical surfaces into productive gardening space.

Choosing the Right Plants for Vertical Gardening

When planning your vertical garden, it's important to choose plants that are well-suited for this style of growing. Plants with *vining or trailing habits* are the best candidates for vertical gardening, as they naturally want to climb or spread. Here are a few options to consider:

- **Climbing Vegetables**: Beans, peas, cucumbers, and tomatoes are great choices for vertical gardens. These plants naturally grow upward and can be trained along a trellis or support system.
- **Herbs**: Thyme, oregano, basil, and parsley do well in small vertical planters or hanging baskets. They don't require much space and can thrive with moderate sunlight.

- **Fruits**: Strawberries are ideal for vertical gardening, especially in stacked planters or towers. They produce small, trailing vines that make harvesting easy. Additionally, dwarf fruit trees can be trained along a trellis, although they require more attention and space.
- **Flowers**: For added beauty, consider integrating flowers like nasturtiums, petunias, or sweet peas into your vertical garden. These flowers can brighten your space while attracting beneficial pollinators like bees and butterflies.

Maintaining Your Vertical Garden

Although vertical gardens are space-efficient, they do require careful attention, particularly in terms of *watering* and *sunlight*. Consistent watering is essential because the soil in vertical planters tends to dry out more quickly than in traditional garden beds. Depending on your setup, you may want to install a **drip irrigation system** that ensures even watering across all levels of your garden. Alternatively, you can manually water your plants, paying extra attention to the containers on the highest levels, as these tend to dry out faster.

Sunlight is another key consideration. Most vegetables and herbs need at least 6 hours of direct sunlight per day. If your vertical garden is mounted on a wall, make sure that the sun reaches all levels evenly. For shaded areas, choose shade-tolerant plants like lettuce, spinach, or mint.

Soil and Fertilization: Just like with container gardening, using high-quality *potting mix* is essential for vertical gardens. Make sure to refresh the soil each growing season to maintain nutrient levels. Fertilize regularly with a balanced organic fertilizer, but be careful not to over-fertilize, as vertical containers are smaller and can't store excess nutrients as well as ground soil.

With the right system, plants, and maintenance practices, you'll be able to maximize your growing potential, even in the smallest spaces.

Next, we'll explore the differences between raised beds and traditional gardening, helping you determine the best approach for your homestead.

Raised Beds vs. Traditional Gardening: Which is Right for You?

When planning your homestead garden, one of the first decisions you'll face is whether to use *raised beds* or opt for *traditional in-ground gardening*. Both methods have their benefits, and the right choice for you depends on your available space, soil conditions, and personal preferences. Let's dive into the pros and cons of each, so you can decide which method will help you maximize your garden's potential.

What Are Raised Beds?

Raised beds are garden plots constructed above ground, usually in a framed wood, metal, or stone structure. The frame is filled with nutrient-rich soil, and plants grow inside this space. Raised beds offer many benefits, particularly for those with poor native soil or limited space, but they also come with some limitations.

Benefits of Raised Beds

- **Better Soil Control**: One of the main advantages of raised beds is that you have complete control over the soil quality. In many parts of the world, native soil can be compacted, lacking in nutrients, or full of rocks, making it challenging to grow healthy plants. You can bring in high-quality soil mixes, compost, and organic matter, ensuring your plants have the best possible environment to thrive. This can be

particularly useful for crops that require rich, well-draining soil, such as carrots or potatoes.
- **Improved Drainage**: Raised beds typically offer superior drainage compared to traditional gardens, which is especially helpful if your garden is prone to waterlogging or heavy rainfall. The soil in raised beds tends to stay looser and better aerated, allowing roots to grow freely and minimizing the risk of root rot.
- **Easier to Maintain**: Raised beds are often elevated to waist height or just above the ground, making them easier to maintain, particularly for those who have mobility issues or prefer not to bend over while gardening. Because the plants are contained in a defined space, raised beds also make it easier to control weeds and prevent soil compaction from foot traffic.
- **Longer Growing Season**: The structures tend to warm up faster in the spring compared to in-ground gardens, giving you a head start on the growing season. The elevation allows the soil to heat up more quickly, which can be especially advantageous in cooler climates where every extra week of growing time counts.

Challenges of Raised Beds

- **Initial Setup Costs**: One of the main drawbacks of raised beds is the initial cost of materials. Building them requires wood, metal, or stone for the frame, as well as high-quality soil and compost to fill the beds. Depending on the size of your garden and the materials you choose, the cost can add up quickly. While this investment will pay off over time, it can be a barrier for beginners on a tight budget.
- **Watering Requirements**: Raised beds generally require more frequent watering than in-ground gardens, as the soil

in them tends to dry out faster, especially in hot weather. You'll need to stay on top of your watering schedule, or consider installing drip irrigation systems or soaker hoses to ensure your plants stay hydrated.

What is Traditional Gardening?

Traditional gardening involves planting directly into the ground, using the soil already present on your property. This is the most common method of gardening and can be incredibly rewarding if you have rich, healthy soil and plenty of space.

Benefits of Traditional Gardening

- **Lower Initial Costs**: Traditional gardening doesn't require the same initial investment as raised beds. You can start planting as soon as you prepare the soil by removing weeds, tilling, and adding compost or other organic matter. If you're looking for a cost-effective option, traditional gardening may be the way to go.
- **Less Frequent Watering**: In-ground gardens often require less frequent watering than raised beds. The soil retains moisture more effectively, which means you won't need to water as often—an advantage in hot, dry climates.
- **More Space for Plants**: Traditional gardening typically allows for more space, which is ideal if you're planting larger crops like corn, squash, or pumpkins that need room to spread out. If you have ample garden space, traditional gardening enables you to grow more without the spatial constraints of raised beds.

- **Utilizing Native Soil**: If your native soil is healthy and rich, traditional gardening allows you to take full advantage of what's already available. This reduces the need to purchase large quantities of soil or amendments, making it more sustainable in the long run.

Challenges of Traditional Gardening

- **Soil Quality and Preparation**: One of the biggest challenges with traditional gardening is dealing with poor soil quality. If your soil is compacted, sandy, or lacking in nutrients, it may take considerable time and effort to amend it. This can involve regular tilling, adding compost, and using fertilizers to improve the soil structure and fertility.
- **Weed Control**: In-ground gardens tend to face more issues with weeds, which can quickly take over if not managed properly. Mulching and regular weeding are essential to keep your garden healthy and productive. Raised beds often reduce the amount of weeding you'll need to do, as the contained space is easier to manage.
- **Physical Labor**: Traditional gardening often requires more physical effort, especially in larger plots. You'll need to bend, kneel, and walk through your garden beds to plant, water, weed, and harvest. While this might be enjoyable for some, it can be physically demanding for others, especially over time.

So, Which Works Best for You?

. . .

When deciding between raised beds and traditional gardening, consider your space, soil quality, budget, and physical capabilities. Raised beds offer more control and can be easier to manage, especially in smaller spaces or for those with mobility issues. On the other hand, traditional gardening is more cost-effective and better suited for larger spaces with healthy soil.

If you're unsure, you can always combine both methods—using raised beds for specific crops that need extra care or better soil conditions while utilizing traditional garden space for hardier plants that can thrive in your native soil.

But there's another innovative method to consider. If you're looking for a solution that eliminates the need for soil altogether and maximizes indoor or small outdoor spaces, hydroponics might be the perfect next step.

Hydroponics for Small Spaces

If you are working with limited space, *hydroponics* offers an alternative, a practical, soil-free method to grow plants indoors or in small outdoor areas. It uses a nutrient-rich water solution to feed plants directly, making it highly efficient in both space and water usage. Whether you're looking to grow leafy greens, herbs, or small vegetables, it is a versatile solution that fits well into urban and small-space gardening.

Why Hydroponics Works for Small Spaces

Hydroponics is perfect for homesteaders with little room because it eliminates the need for large garden beds and allows plants to grow closer together. Without soil, plants rely on the nutrient solution for growth, resulting in faster growth rates and higher yields per square foot. You can even stack hydroponic systems vertically to make the most of small spaces, maximizing your growing area indoors or on a small balcony.

Simple Hydroponic Systems for Beginners

If you're new to hydroponics, there are two easy-to-set-up systems that work well in small spaces:

- **Deep Water Culture (DWC)**: This is one of the simplest and most popular choices for beginners. Plants are suspended in *net pots* with their roots submerged directly in nutrient-enriched water. An *air pump* provides oxygen to the roots, preventing them from drowning and keeping the water aerated. DWC systems are easy to build at home using a large container, a water pump, and net pots to hold the plants. Leafy greens like lettuce and spinach thrive in this system due to their fast-growing nature and shallow root systems.
- **Wick System**: This is another beginner-friendly option that requires no pumps or electricity. It works by placing plants in a growing medium such as *coconut coir* or *perlite*, while a wick (made from cotton or nylon) draws the nutrient solution from a reservoir to the plant's roots. This passive system is perfect for small-scale indoor gardening, especially for plants like herbs and small peppers, that don't require high levels of water or nutrients.

Starting Your Hydroponic Setup

Setting up a basic hydroponic system is straightforward.

- Start with plants that thrive in hydroponic systems and are easy to grow, such as lettuce, herbs, spinach, and strawberries.
- Find a location with plenty of *natural light* or install *grow lights* for indoor setups. Plants typically need *12-16 hours of light* per day to thrive in a hydroponic system. Small spaces like kitchen counters, balconies, or sunny windowsills work well.
- Fill the reservoir with water and add a hydroponic *nutrient solution* according to product instructions. Most kits include easy-to-use mixes that provide all the necessary nutrients for plant growth.
- Monitor *pH levels* to maintain the right balance. Plants generally thrive at a pH between *5.5 and 6.5*. Use a simple pH test kit to check your water and adjust as needed with pH up or down solutions.

Now that you've also explored hydroponics, it's time to choose the best approach for your unique space and needs.

Bringing It All Together: Choosing Your Gardening Method

You might be wondering how these different gardening methods fit together and where to begin. Each one has its own benefits depending on your space and soil, but remember—they don't have to exist in isolation. Here's a simple guide to help you choose based on your circumstances:

- **Limited space?** Start with *containers* and incorporate *vertical gardening* to maximize your growing area. If you're looking for an innovative option, you might also consider *hydroponics*, which allows you to grow plants more efficiently with minimal space.

- **Poor soil?** *Raised beds* offer control over your soil quality, allowing you to create the perfect mix for healthy plant growth. *Hydroponics* is another solution, bypassing the need for soil entirely while offering an alternative for plants that need special care.
- **More land?** Use *traditional gardening* for larger crops, while saving *raised beds* for plants that require extra attention.

Remember, the best approach is to begin with a manageable scale and expand as you learn. It's time to think about how to do it cost-effectively. In the next section, we'll explore practical ways to start your garden on a budget options that ensure you don't break the bank while pursuing self-sufficiency.

Gardening on a Budget: DIY Tools and Affordable Seeds

Homesteading doesn't have to be expensive, especially when it comes to gardening. In fact, one of the joys of it is finding creative ways to work with what you have and using affordable resources to build something sustainable. Gardening on a budget is a smart way to start, allowing you to gradually grow your homestead without overwhelming your finances.

Starting with Affordable Seeds

Seeds are the heart of any garden, but that doesn't mean you need to spend a lot of money on them. Here are a few ways to find affordable seeds:

- **Seed Swaps and Local Seed Libraries**: Many communities have seed swap events where gardeners exchange seeds they've saved from their own gardens. These gatherings are a great way to get a variety of seeds for free while connecting with other gardeners in your area. Additionally, some libraries and community centers offer seed libraries, where you can "borrow" seeds for free and return seeds you've harvested at the end of the season.

- **Heirloom and Open-Pollinated Seeds**: Investing in heirloom or open-pollinated seeds can save you money in the long run because you can *save seeds* from your harvest to plant the following year. These types of seeds produce plants that grow true to their type, meaning you'll be able to grow the same quality crops year after year without buying new seeds.
- **Bulk Seeds**: If you're planning a larger garden, consider buying seeds in bulk. Some companies sell seeds by the pound, which can be much cheaper than buying individual seed packets. This is particularly helpful l for crops like lettuce, beans, or radishes, which can be grown in large quantities.
- **Grow from Scraps**: Another budget-friendly option is to *regrow vegetables from kitchen scraps*. Vegetables like green onions, lettuce, celery, and even potatoes can be regrown from the parts you'd normally discard. Simply place the scrap in water or soil, and within a few weeks, you'll have a new plant.

DIY Garden Tools and Structures

When starting a garden, you don't need to invest in expensive tools or garden structures. Many essential gardening tools and materials can be made from items you likely already have at home. Here are a few budget-friendly DIY ideas:

- **DIY Raised Beds**: Raised beds don't have to be expensive to build. You can create simple, sturdy raised beds from *reclaimed wood*, old pallets (make sure they are safe for gardening use), or even bricks or cinder blocks. Repurposing materials saves money and reduces waste, making it a sustainable option for your garden.
- **Homemade Compost Bin**: Composting is one of the best ways to create nutrient-rich soil for your garden without spending money on store-bought fertilizers. A simple compost bin can be made from old pallets, chicken wire, or even a plastic storage container with holes drilled in it for airflow. As you add food scraps, grass clippings, and other organic matter, you'll create a steady supply of compost to enrich your soil.
- **Repurposed Planters**: You don't need to buy fancy planters to start your garden. Many household items can be repurposed as containers for your plants. Old buckets, tubs, or even broken pieces of furniture can be turned into functional and decorative planters. Just make sure to drill drainage holes in the bottom to prevent water from pooling and causing root rot.
- **DIY Trellises and Supports**: For vertical gardening or growing climbing plants like beans or cucumbers, you can easily make your own trellises from materials like bamboo poles, old fencing, or twine. You can even use sticks or branches from your yard to create a simple teepee structure for your climbing plants. This low-cost, functional solution can support your plants without the need for expensive store-bought options.
- **Watering Systems**: Efficient watering is key to keeping your garden thriving, but you don't need an expensive irrigation system to do the job. Consider installing a DIY drip irrigation system using inexpensive materials like a hose, emitters, and connectors. For an even simpler solution, you can create a *self-watering system* by repurposing plastic bottles. Poke small holes in the bottle, bury it in the

soil next to your plants, and fill it with water—this will provide slow, steady moisture directly to the roots.

Using Mulch and Natural Fertilizers

Maintaining a healthy garden doesn't have to involve expensive fertilizers or soil conditioners. Here are a few budget-friendly ways to keep your soil healthy and your plants thriving:

- **Mulch from Yard Waste**: Instead of buying mulch from a garden center, use yard waste like grass clippings, leaves, or wood chips. It helps retain moisture, suppress weeds, and add nutrients to the soil as it breaks down.
- **Homemade Fertilizer**: You can make your own organic fertilizer using materials like coffee grounds, eggshells, or banana peels. For example, *compost tea* made from compost or worm castings can be a great liquid fertilizer for your plants. Simply steep the compost in water for a few days and then use the liquid to feed your plants.
- **Manure**: If you have access to livestock or know someone who does, manure is an excellent, free source of nutrients for your garden. Be sure to compost the manure for several months before applying it to your garden to avoid burning your plants.

Gardening on a budget is possible and rewarding; however, as your garden grows and your needs evolve, you might realize this is just part of a more comprehensive homesteading experience. Raising small farm animals can be an interesting next step if you aim for greater self-sufficiency and sustainability. In the next chapter, we'll explore how to get started with small livestock, helping you move toward a more balanced, thriving homestead.

4
RAISING SMALL LIVESTOCK FOR BEGINNERS

Best Livestock for Small Homesteads: Chickens, Rabbits, Bees

Once you've established your garden, adding small livestock to your homestead is a great way to take your self-sufficiency to the next level. Before deciding which one is right for your situation, it's essential to check if you or anyone in your household has **allergies** that might be triggered by the animals you're considering. For example, chickens can produce dander that may cause respiratory allergies, rabbits shed fur and dander, and beekeeping involves the risk of **bee stings**, which can cause severe allergic reactions. If you're uncertain about potential allergies, consult a healthcare provider before committing to raising any livestock.

Once you've considered any potential health concerns, you can confidently explore the best options for you. Small animals like chickens, rabbits, and bees are ideal for beginners, as they require less space, are relatively easy to care for, and offer a range of benefits—from fresh eggs and meat to honey and natural fertilizers. Let's explore these three options to help you decide which one (or more) might best fit your homestead.

Chickens: A Homestead Staple

Chickens are often the first livestock choice for beginner homesteaders. They are relatively low-maintenance and provide a consistent supply of eggs and meat if you choose to raise dual-purpose breeds.

Why Chickens?

- **Egg Production**: Hens can lay an egg almost every day during their peak laying years, giving you a steady supply of fresh, homegrown eggs. Most breeds start laying around 5 to 6 months of age, and you can expect consistent egg production for at least a few years.
- **Meat Production**: If you're interested in raising chickens for meat, there are dual-purpose breeds like Rhode Island Reds or Plymouth Rocks that are good for both eggs and meat. Alternatively, meat-specific breeds like Cornish Cross chickens can be raised for a few months before processing.
- **Natural Fertilizer**: Chickens produce excellent manure that can be composted and used to enrich your garden soil. They're also natural pest controllers, eating bugs and weeds in your garden.

Space Requirements

. . .

Chickens don't need a lot of space, but they do need a secure coop and outdoor run to protect them from predators. A good rule of thumb is to provide 2-3 square feet of indoor space per chicken in the coop and 8-10 square feet per chicken in the outdoor run. If you have a smaller backyard, consider a **chicken tractor**, a moveable coop that allows you to rotate the chickens around your yard, fertilizing the soil as they go.

Challenges

One thing to be aware of with chickens is the **noise**. Roosters, in particular, can be loud, so if you're in a suburban area with close neighbors, you might want to stick with hens only. Additionally, while chickens are hardy, they can be susceptible to diseases like mites and parasites, so regular health checks and clean living conditions are essential.

Rabbits: Quiet and Efficient Meat Producers

Rabbits are an excellent option for small homesteads, particularly if you're interested in raising animals for meat. They require very little space, are quiet, and reproduce quickly, making them a highly efficient source of protein.

Why Rabbits?

- **Meat Production**: Rabbit meat is lean, high in protein, and often touted as one of the most efficient meats to raise on a small scale. A single breeding pair can produce several

litters a year, and rabbits grow quickly—reaching processing weight in as little as 10-12 weeks.
- **Low Space Requirements**: Rabbits can be kept in small hutches, cages, or rabbit tractors that allow them to graze in your yard. They don't require much space, making them ideal for suburban homesteads.
- **Fertilizer**: Rabbit manure is an excellent fertilizer that doesn't need to be composted before being applied to your garden. It's rich in nitrogen and phosphorus, making it a valuable addition to your soil.

Space Requirements

Rabbits need a clean, dry space to live, with room to move around and access to fresh water and food. A basic rabbit hutch or cage should provide at least 3-4 square feet of space per rabbit. Many homesteaders use **rabbit tractors**—similar to chicken tractors—so the rabbits can graze on grass and fertilize the ground as they move around.

Challenges

While rabbits are relatively low-maintenance, they do require regular care and attention. Heat can be dangerous for rabbits, so you'll need to ensure they have shade and proper ventilation during the hotter months. Additionally, since rabbits are prey animals, their enclosures need to be predator-proof.

Bees: Pollinators with Sweet Rewards

. . .

Raising bees can be an incredibly rewarding addition to any homestead. Not only do they produce honey, but they also play a critical role in pollinating your garden, which can increase your crop yields. Bees are a great choice if you want to promote a healthy ecosystem.

Why Bees?

- **Honey Production**: Bees produce honey, a natural sweetener that's great for cooking, preserving, and even medicinal use. A healthy hive can produce up to 60 pounds of honey per year.
- **Pollination**: Bees are essential pollinators, helping to improve the health and yield of your garden. If you're growing fruits and vegetables, bees can significantly boost your crop production.
- **Beeswax and Other Byproducts**: In addition to honey, bees produce wax that can be used for candles, cosmetics, and other household products.

Space Requirements

Bees require very little space compared to other livestock. A couple of hives can fit in even the smallest backyards as long as the bees have access to flowering plants for foraging. You'll need to provide a suitable location for the hives, preferably in a sunny spot sheltered from the wind.

Challenges

. . .

Beekeeping requires some initial equipment investment, including bee suits, smokers, and hives. Additionally, while bees generally take care of themselves, you'll need to monitor the health of your hives and manage them to prevent swarming or disease outbreaks.

Choosing the Right Livestock for You

Each of these animals offers unique benefits for small homesteads. The best choice depends on your available space, time, and what you hope to gain from raising livestock. Whether you start with just one type or decide to mix and match, small livestock can greatly enhance your homesteading experience by providing fresh food and improving your garden's sustainability.

In the next section, we'll explore the legal considerations and space requirements to help you ensure that your homestead is compliant with local laws and set up for success.

Legal Considerations and Space Requirements

Before bringing any livestock onto your homestead, it's essential to consider the legal aspects and ensure you have enough space to care for your animals properly. Depending on where you live, specific laws and zoning regulations may dictate what types of animals you can raise and how many you're allowed to keep. Let's walk through the key legal concepts you need to evaluate.

Zoning Laws and Regulations

Zoning laws can vary widely depending on your location—urban, suburban, or rural—and these regulations often dictate what types of livestock are permitted. Urban areas are generally more restrictive, while rural areas typically allow for more flexibility in raising animals. Here are some important points to keep in mind:

- **Check Local Zoning Laws**: The first step is to check with your local government or city planning office. They will be able to inform you about restrictions on livestock, such as limits on the number of animals you can keep, requirements for how far enclosures need to be from property lines, and whether certain animals (like roosters or bees) are allowed in your area.
- **Permits and Licenses**: Some areas require you to obtain a permit or license to keep certain types of livestock. For example, keeping bees in urban areas often requires a beekeeping permit to ensure that the hives are placed safely and do not pose a risk to neighbors. Similarly, if you want to raise chickens, some municipalities may limit the number of hens or ban roosters altogether to reduce noise complaints.
- **Homeowners' Association (HOA) Rules**: If you live in a community governed by an HOA, be sure to check their bylaws as well. Many HOAs have strict rules regarding pets and livestock, even in suburban areas where local laws might be more lenient. HOAs can enforce restrictions on fencing, noise, and even the size of your livestock enclosures.
- **Noise and Nuisance Laws**: Even if your local laws allow for livestock, you'll need to be mindful of noise ordinances and nuisance laws, particularly if you live in a suburban or urban area. Roosters, for example, can be quite noisy and are often prohibited within city limits. Keeping bees might also require you to place hives a certain distance from neighbors to prevent potential stings or swarming.

Space Requirements

. . .

Once you've navigated the legal considerations, it's time to assess whether you have enough space to provide a healthy environment for your animals. Each type of livestock comes with its own space needs, and it's important to make sure you can meet them for the health and well-being of your animals.

- **Chickens** require both *indoor* and *outdoor space* to thrive. Inside the coop, each chicken needs around **2-3 square feet** of space, while their outdoor run should offer **8-10 square feet per chicken**. If you're working with limited space, a chicken tractor can be a great way to give your chickens access to fresh ground while keeping them safe from predators. Just make sure to provide a safe, secure area where they can roost and lay eggs.
- **Rabbits** are ideal for small homesteads because they don't need much room. A basic hutch or cage should provide at least **3-4 square feet per rabbit**, and you can supplement this with a rabbit tractor or an outdoor pen that allows them to graze safely. It's important that their enclosure offers protection from the elements and predators and that they have enough room to move, hop, and stretch.
- **Bees** are the easiest livestock to care for in terms of space. You only need a small area for a couple of hives, but placement is key. Bees need access to flowering plants, and their hives should be placed in a sunny, sheltered spot that is far enough away from neighbors or walkways. Depending on your local regulations, you might need to keep hives a certain distance from property lines or provide barriers, such as fences, to direct bee flight paths upward.

Navigating Legal and Space Requirements

. . .

By understanding the legal requirements and ensuring you have the appropriate space for your animals, you can avoid potential issues down the road. Always check with local authorities and, when necessary, obtain the proper permits before adding livestock to your homestead. With these preparations in place, you can confidently move forward with raising animals, knowing that both you and your livestock are set up for success.

Basic Care: Feeding, Housing, and Maintenance

Once you've navigated the legal aspects and prepared your space, the next crucial step is ensuring you understand the basic care, feeding, housing, and maintenance required for your small livestock. Different animals have unique needs, but caring for them with the right setup and routine becomes a rewarding part of your homesteading journey. Let's explore the fundamental requirements for each type of animal to keep them healthy and thriving.

Chickens: Basic Care and Feeding

Chickens are relatively easy to care for, but they have specific needs for feeding, housing, and general maintenance.

- **Feeding**: Chickens require a balanced diet to stay healthy and produce eggs consistently. A high-quality layer feed is essential for hens, providing the nutrients they need to produce strong eggshells and maintain good health. You can supplement their diet with kitchen scraps, fresh greens, and grains, but be careful not to feed them anything harmful, like onions, avocados, or chocolate. Chickens are natural foragers and will happily eat insects and weeds in your yard, which also helps control pests.

- **Housing**: A secure and well-ventilated **coop** is critical for protecting chickens from predators and the elements. Inside the coop, provide nesting boxes filled with straw or wood shavings where hens can lay their eggs. Chickens also enjoy dust baths, which help keep them clean and free from mites, so consider providing an area with loose dirt or sand.
- **Maintenance**: To keep your flock healthy, you'll need to clean the coop regularly, removing old bedding, droppings, and uneaten food. Fresh water should always be available, and feeders and waterers should be cleaned weekly to prevent disease. Check for signs of parasites or illness, and ensure the coop remains secure to keep out predators like foxes, raccoons, or hawks.

Rabbits: Basic Care and Feeding

Rabbits are quiet and efficient but need proper housing and a nutritious diet to stay healthy.

- **Feeding**: The foundation of a rabbit's diet is **hay**, which should make up the majority of their food intake. Hay keeps their digestive system functioning properly and helps wear down their teeth, which grow continuously. Along with hay, rabbits should be given a limited amount of high-quality rabbit pellets and fresh vegetables like leafy greens, carrots, and celery. Avoid giving rabbits sugary treats or foods that could upset their digestive systems, such as iceberg lettuce.
- **Housing**: Rabbits require a clean, dry, and secure hutch or cage with enough space for them to move, hop, and stretch out. Ensure the enclosure protects from the elements, especially during extreme heat or cold. As previously mentioned, a rabbit tractor, a moveable hutch that allows

rabbits to graze, can be an excellent option for rotating them around your yard.
- **Maintenance**: Regularly clean the rabbit hutch or cage to prevent waste buildup, which can attract flies and lead to health problems. Fresh water should always be available, and check your rabbits for signs of illness or overgrown teeth. Groom them occasionally to keep their fur in good condition and ensure their claws don't become too long.

Bees: Basic Care and Maintenance

While bees mostly take care of themselves, they still need regular monitoring to ensure their hives remain healthy and productive.

- **Feeding**: Bees forage for nectar and pollen, but they may need supplemental feeding during periods of drought or when fewer flowers are available. Beekeepers often provide sugar water (a mixture of water and sugar) to ensure the bees have enough energy to survive until more flowers bloom.
- **Housing**: Bees live in **hives**, which need to be kept in a sunny, sheltered spot away from strong winds. You'll need to set up a hive box, frames, and other equipment like a bee suit and smoker to inspect the hive. It's essential to position the hive where the bees have easy access to flowering plants for foraging.
- **Maintenance**: Bees generally require less hands-on care than chickens or rabbits, but you'll need to inspect the hive regularly to check for diseases, pests (like varroa mites), and signs of swarming. Cleaning the hive periodically and providing proper ventilation is also essential. As a beekeeper, you must ensure your bees have enough honey stored to survive the winter or supplement it with sugar water if necessary.

Simple, Low-Cost Livestock Housing Solutions

Providing proper housing for your livestock doesn't have to be expensive. Many affordable, DIY options exist to create safe and functional animal shelters. The key is to focus on building structures that meet the animals' basic needs—protection from the elements, security from predators, and adequate ventilation—while using materials that are either low-cost or easily accessible.

Chickens: Affordable Coop Solutions

A well-designed chicken coop is essential for keeping your flock safe and healthy, but that doesn't mean it needs to be expensive. Here are a few low-cost options for building a coop:

- **Recycled Materials**: One of the easiest ways to save money on a chicken coop is to use reclaimed wood, old pallets, or even an old shed as the base structure. As long as the materials are sturdy and weatherproof, you can convert them into a secure coop with a little bit of work. Make sure to add proper ventilation and roosting bars for the chickens to sleep on.
- **Chicken Tractor**: If you have a smaller backyard, consider building a chicken tractor, which we've already mentioned is a mobile coop that can be moved around your yard. This allows your chickens to forage for fresh grass and bugs, naturally fertilizing your soil as they go. You can build a basic chicken tractor using inexpensive materials like PVC pipe, wire mesh, and lightweight wood.
- **Roofing and Insulation**: To keep your chickens safe from the weather, you'll need to add a simple roof. Corrugated

plastic or metal sheets are both inexpensive options. Insulation isn't always necessary, but in colder climates, you can add straw or old blankets to the coop to help keep your chickens warm during winter.

Rabbits: Simple Hutch and Pen Ideas

Rabbits require safe, dry housing with enough space to move and stretch. Fortunately, building a rabbit hutch can be a low-cost project with the right materials.

- **DIY Rabbit Hutch**: A basic rabbit hutch can be built from reclaimed wood or pallets and covered with wire mesh to create a secure space for your rabbits. The hutch should be elevated off the ground to protect the rabbits from moisture and predators. Make sure the hutch includes a nesting area where rabbits can hide and rest.
- **Rabbit Tractor**: Similar to a chicken tractor, a rabbit tractor is a mobile pen that allows your rabbits to graze on fresh grass. It can be made from lightweight wood or PVC pipe and covered with wire mesh to keep the rabbits safe. This type of housing provides natural food for your rabbits while allowing you to rotate them to different parts of your yard.
- **Insulation and Shade**: Rabbits are sensitive to heat, so it's important to provide shade during hot months. You can use old sheets, tarps, or shade cloth to create a cooler environment for them. Adding straw or old blankets inside the hutch in winter can help keep the rabbits warm.

Bees: Low-Cost Hive Solutions

HOMESTEADING ESSENTIALS

. . .

Beekeeping requires specific equipment, but you can still save money by building your own hive and using simple solutions for your beekeeping setup.

- **DIY Top-Bar Hive**: A top-bar hive is a simple, low-cost alternative to traditional box hives. It can be made from basic materials like untreated wood and requires fewer accessories than a standard Langstroth hive. Top-bar hives are easier for beginners to manage and allow bees to build natural combs, which reduces the cost of buying frames.
- **Recycled Materials for Hive Stands**: Instead of purchasing expensive hive stands, you can use cinder blocks, bricks, or wooden crates to elevate your hives off the ground. This protects the hive from moisture and keeps it at a comfortable working height for you.
- **Beehive Placement**: Proper placement can also save you time and money. Make sure to position your hives in a sunny, sheltered spot that offers protection from strong winds. This will help keep your bees healthy and reduce the need for extra insulation or hive management in extreme weather.

By using recycled materials and simple designs, you can create functional, low-cost housing for your livestock without sacrificing their health or safety. A little creativity and resourcefulness can go a long way. In the next chapter, we'll dive into the importance of sustainability on your homestead, looking at ways to make it more self-sufficient while keeping costs low.

SUSTAINABILITY ON A BUDGET

Planning a Low-Cost Homestead: Saving Without Sacrificing Quality

Building a homestead doesn't have to drain your finances. In fact, one of the core values of homesteading is **self-sufficiency**, which includes finding ways to create a sustainable lifestyle without overspending. While it's easy to be tempted by shiny new tools, expensive materials, and ready-made solutions, there are plenty of ways to **save money** while still achieving a high-quality, productive homestead. The key is to plan carefully, prioritize your investments, and look for creative solutions that reduce costs without sacrificing efficiency or quality.

Start with a Budget and Prioritize

The first step to creating a low-cost homestead is to *set a budget* and decide what's most important. Break your goals into short-term and long-term plans. For example, in the first year, your focus might be on establishing a garden and raising small livestock like chickens. You might aim to add more livestock or install renewable energy systems in the long term.

Once your goals are clear, prioritize *where to spend your money*. Some initial investments, like good-quality tools or building materials for livestock shelters, may be necessary upfront. However, other projects, like expanding your garden or adding compost systems, can be done gradually over time, allowing you to spread out the costs. By focusing on what's essential now and what can wait, you can avoid the temptation to overspend early on.

Utilize Free or Recycled Materials

One of the best ways to save money is to *reuse materials* you already have or find items for free or at low cost. Look for building supplies at construction sites, local classifieds, or community groups, where items like wood, pallets, bricks, or fencing components are often available for free or at a steep discount. Many homesteaders build raised beds, coops, and compost bins from reclaimed wood, making these structures both cost-effective and sustainable.

You can also find *second-hand tools* at yard sales or through online marketplaces, often at a fraction of the cost of new ones. If well-maintained, they can last just as long as their brand-new counterparts. For more specialized equipment, consider *borrowing or renting* before making a purchase, especially if it's something you won't use often.

DIY Projects

Do-it-yourself projects are a cornerstone of low-cost homesteading. Many projects can be done with basic skills and minimal materials, from building your own raised garden beds to creating a simple rainwater collection system. Not only does this save money, but it also allows you to customize your setup to fit your specific needs.

Start with small, manageable projects that align with your skill level, and as you gain confidence, tackle more complex builds. You'll find that DIY solutions can provide the same functionality as store-bought options at a fraction of the cost. Additionally, online resources and homesteading communities are great places to find tutorials and guides for building anything from a chicken coop to solar-powered lighting systems.

Grow Your Own Food and Share with Others

Needless to say, one of the simplest ways to save money is to grow your own food. Whether you're starting with a few containers on a balcony or planning a large garden, growing your own vegetables, herbs, and fruits is far more cost-effective than buying them. As your garden expands, consider sharing surplus produce with neighbors or bartering with other homesteaders for items you may need but don't have the space or skills to produce yourself.

In the next section, we'll explore specific DIY projects you can start with to reduce costs further and increase your self-sufficiency.

DIY Projects: Building a Compost Bin, Rainwater Collection Systems, and More

A significant part of saving money while homesteading is taking on *do-it-yourself projects*. This way, you'll reduce your upfront costs, and you will be able to tailor your homestead to meet your specific needs. Two of the most useful and cost-effective DIY projects you can start with are building a *compost bin* and setting up a *rainwater collection system*. These projects promote sustainability, help you recycle natural resources, and contribute to the overall efficiency of your homestead.

DIY 1: BUILDING A COMPOST BIN

Composting is one of the simplest and most beneficial practices for any homestead. It reduces waste and provides nutrient-rich compost for your garden. Fortunately, compost bins can be built inexpensively using repurposed or low-cost materials.

- **Materials**: The most common materials used for compost bins are reclaimed wood, pallets, or wire mesh. Wooden pallets, in particular, are often available for free or at low cost and make excellent frames for a compost bin. You can also use chicken wire or hardware cloth to create an enclosure that allows airflow while keeping your compost pile contained.
- **Construction**: To build a simple compost bin, create a three-sided structure using pallets or wooden slats and leave the front open or use a removable panel for easy access. Ensure there's enough space for the compost to break down, but not so much that it becomes unmanageable. A bin that's about 3x3 feet works well for most homesteads. Adding a hinged lid or tarp can help keep the compost covered and protected from excessive rain or drying out in the sun.
- **Usage**: Once your compost bin is ready, fill it with kitchen scraps, yard clippings, and organic materials like coffee grounds or egg shells. Regularly turning the pile will speed up decomposition, and in just a few months, you'll have nutrient-rich compost to add to your garden beds.

For those interested in experimenting with different designs, here are a few alternative compost bin options that can be customized to suit your homestead's unique needs.

Insight: Exploring Different Designs and Options

While the basic compost bin design is simple and effective, you can customize it to suit your specific space, aesthetic preferences, or composting needs. Let's explore a few alternative designs and maintenance tips that will help you create a composting system that works best for your homestead.

1. Recycled Plastic Barrel Compost Bin

For homesteaders with limited space or those who want a more compact and portable option, using a *recycled plastic barrel* to create a compost bin is a great solution. These barrels are easy to find and repurpose and are perfect for small homesteads or urban settings.

- **Materials Needed**: A 55-gallon plastic barrel, a drill, and a secure lid.
- **Construction**: Drill holes in the sides of the barrel to allow for airflow. Place the barrel on a stand, such as cinder blocks, to keep it elevated. A lid will help contain the compost and prevent pests.
- **Usage**: This type of bin is perfect for smaller amounts of kitchen waste and garden clippings. Simply roll the barrel occasionally to mix the compost, ensuring even decomposition.

2. Multi-Bin Compost System for Large Homesteads

Consider building a *multi-bin system* if you have more space and a larger volume of compostable materials. This allows you to have different stages of compost at once, speeding up the process by letting one bin decompose while filling another.

- **Materials Needed**: Reclaimed wood or pallets, wire mesh, and hinges for removable panels.
- **Construction**: Build three bins side by side, with removable front panels for easy access. As one bin fills up, you can start a new batch in the second bin while the first continues to break down.
- **Usage**: This method is ideal for larger homesteads that produce more waste. It offers a continuous supply of compost and helps manage larger quantities efficiently.

3. Tumbler Composters: Faster Decomposition

For those looking for a quicker way to compost, *tumbler composters* are a great option. These rotating bins speed up the composting process by making it easier to turn the pile regularly.

- **Materials Needed**: A purchased tumbler composter or a DIY drum with a rotating mechanism.
- **Construction**: If making your own, you can mount a plastic or metal drum on a sturdy frame with a handle or mechanism that allows it to rotate.
- **Usage**: Simply rotate the composter every few days to aerate the compost. Tumblers are particularly useful for people with limited physical ability, as they require less effort than turning a traditional pile by hand.

4. Vermicomposting: A Solution for Small Spaces

. . .

If you work with very limited space, *vermicomposting* (composting with worms) is an excellent alternative. This method is particularly useful in urban environments where outdoor space is scarce.

- **Materials Needed**: A small bin, red wiggler worms, and organic materials such as vegetable scraps and newspaper.
- **Construction**: Create a bin with small drainage holes, fill it with moist bedding material (like shredded newspaper), and add worms. Keep it indoors or in a sheltered area.
- **Usage**: Vermicomposting is ideal for small-scale composting, especially in apartments or small homes. The worms break down food scraps into rich compost, perfect for container gardening.

Maintaining Your Compost Bin: Tips for Success

No matter which design you choose, proper maintenance is key to producing high-quality compost. Here are a few essential tips:

- **Moisture Control**: Your compost should feel like a damp sponge—too dry, and decomposition will slow down; too wet, and it may start to smell. Add water when necessary or dry materials like straw if it's too wet.
- **Airflow**: Turn or aerate the compost regularly to introduce oxygen, which helps break down materials faster. If using a static bin, turning it with a pitchfork every couple of weeks will help.
- **Balancing Green and Brown Materials**: To speed up decomposition and prevent odors, maintain a balance of *green materials* (like food scraps and grass clippings) and *brown materials* (like dried leaves, cardboard, or straw).

HOMESTEADING ESSENTIALS

DYI 2: SETTING UP A RAINWATER COLLECTION SYSTEM

A rainwater collection system is an excellent way to conserve water and reduce your reliance on municipal supplies. With just a few simple materials, you can capture and store rainwater for use in your garden or for other homestead needs.

- **Materials**: To build a basic rainwater collection system, you'll need a rain barrel (which can be an old food-grade drum or plastic barrel), a gutter diverter, and a spigot to access the collected water easily. Rain barrels can often be found at second-hand stores, or you can repurpose containers you already have.
- **Construction**: Position the rain barrel under a downspout to catch water flowing from your roof. You can install a simple diverter in your gutter system to direct rainwater into the barrel while preventing overflow. Adding a spigot near the bottom of the barrel makes it easy to attach a hose or fill watering cans. Be sure to place a mesh screen over the top of the barrel to keep debris and insects out of the water.
- **Usage**: Rainwater collected in barrels can be used to water your garden, wash outdoor tools, or for other non-potable uses around the homestead. This not only reduces your water bill but also provides a sustainable solution for your homesteading needs.

If you want to customize or enhance your rainwater collection setup, here are a few advanced options to consider for increased capacity, efficiency, and expanded use.

Insight: Advanced Options for Efficiency

. . .

While a basic rainwater collection system is simple to set up, you can customize it to suit your specific needs in several ways. Whether you're looking for a more sustainable setup or need a system that handles larger volumes of water, these ideas will help you make the most of your rainwater collection efforts.

1. Linking Multiple Barrels for Increased Capacity

If your homestead requires more water for gardening or livestock, linking multiple barrels together can increase your system's capacity while still using the same downspout for collection.

- **Materials Needed**: Two or more rain barrels, PVC pipe or hose, a spigot, and a connector kit.
- **Construction**: After setting up your first rain barrel under the downspout, install a connector near the top of the barrel that links to a second (or more) barrel. As the first barrel fills, water will flow into the second barrel, allowing you to collect more water without additional downspouts.
- **Usage**: This system is ideal for homesteads that need a higher volume of water, particularly during dry seasons. Multiple barrels ensure that even heavy rains are fully captured.

2. Gravity-Fed Irrigation System

To make watering your garden even easier, you can modify your rainwater collection system to work as a *gravity-fed irrigation system*, eliminating the need to water your plants manually.

- **Materials Needed**: Rain barrel, spigot, hose, and an irrigation kit (drip lines or soaker hoses).
- **Construction**: Place your rain barrel on an elevated platform to create enough gravity pressure to feed water through a hose or irrigation lines. Attach a hose to the spigot and run it through a drip irrigation system or soaker hoses laid out in your garden beds.
- **Usage**: This setup allows the water to flow directly to your plants without additional effort, conserving water and saving time in the process. It's especially useful for vegetable gardens or flower beds that require consistent watering.

3. Rainwater Filtration for Expanded Uses

While the rainwater collected is primarily for non-potable uses like watering the garden, washing tools, or filling livestock troughs, adding a *simple filtration system* can expand the water's usefulness around your homestead.

- **Materials Needed**: A filter designed for rainwater collection, activated charcoal, and a fine mesh screen.
- **Construction**: Install a basic rainwater filtration system between the downspout diverter and the rain barrel, or attach it to the barrel's output spigot. A filter with activated charcoal can remove impurities, making the water cleaner for use in greywater systems or washing outdoor surfaces.
- **Usage**: Filtered rainwater is still not suitable for drinking without further treatment, but it's much cleaner for household tasks like washing outdoor tools, equipment, or even your homestead structures.

Maintenance Tips for Long-Term Efficiency

Regular maintenance is essential to keep your rainwater collection system functioning effectively over time:

- **Check for Leaks**: Ensure that connectors, spigots, and seals remain tight. Any leaks will reduce the amount of water you collect, especially in systems with multiple barrels.
- **Clean the Filter and Barrel**: If your system includes a filter, clean or replace it regularly to prevent blockages. Clean the barrel once a year to remove any sediment buildup that may affect water quality.
- **Protect from Freezing**: In colder climates, disconnect the system during winter months to prevent the water in the barrel from freezing and cracking. Alternatively, drain the barrel or install a frost-proof system.

Other DIY Projects

There are countless other DIY projects that can enhance your homestead's efficiency and sustainability. For instance, you can construct a **DIY wind turbine** to generate renewable energy for your home. Wind turbines are an excellent project for those in areas with consistent wind, helping reduce reliance on the grid and tap into natural energy resources.

Additionally, creating **natural insect repellents from herbs** grown on your homestead is a simple and effective way to protect your garden. Plants like lavender, mint, and lemongrass can be used to make homemade sprays that are free of harsh chemicals but still highly effective at keeping pests away from your crops.

For year-round gardening, building **cold frames for winter**

gardening allows you to extend the growing season and protect delicate plants during colder months. A cold frame, made from old windows or plastic, acts like a mini greenhouse, trapping heat and shielding plants from frost. It's an easy way to keep your garden productive even in winter.

By experimenting with these DIY projects with a little creativity and resourcefulness, the possibilities are endless. By taking on these projects, you save money and contribute to a more sustainable, resilient homestead. In the next section, we'll explore ways to use thrift store finds and other low-cost resources to further reduce your expenses.

Thrift Store Hacks for Homesteaders

Thrift stores, yard sales, and online marketplaces are treasure troves for homesteaders, offering a wide range of affordable items that can be used for gardening, livestock care, and DIY projects. With a bit of ingenuity, these second-hand finds can be transformed into valuable resources for your homestead, saving money while promoting sustainability.

Affordable Gardening Tools and Containers

Gardening tools and supplies can quickly become expensive if purchased new, but thrift stores often carry *used tools* that are still in good condition. Look for hand tools like trowels, pruners, and rakes, which can often be found for a fraction of their original cost. *Old wheelbarrows* or *buckets* are also useful for transporting soil or compost around the garden.

- **Repurposed Containers**: Thrift stores are full of inexpensive containers that can be repurposed for your

garden. Old ceramic pots, plastic bins, and even baskets can make excellent planters for herbs or small vegetables. You can also find large plastic tubs or barrels, which can be turned into raised garden beds or used as water containers for rainwater collection systems.
- **Vintage Items for Garden Décor**: Many homesteaders enjoy adding a touch of charm to their gardens with vintage items found at thrift stores. Old wooden crates, metal watering cans, or cast-iron garden decorations can add both style and functionality to your outdoor spaces.

Upcycled Furniture and Structures

Thrift stores are excellent places to find furniture and materials that can be repurposed for homestead projects.

Old **wooden bookshelves** or **bed frames** can be disassembled and repurposed into raised garden beds or **storage shelves** for your barn or garden shed. These materials are often much cheaper than buying new wood, and with a bit of creativity, they can be adapted to fit your needs.

You can also repurpose **old furniture** to create shelters for your livestock. For example, a discarded wooden dresser can be turned into a **rabbit hutch** or a brooding box for chicks. Add a wire mesh for ventilation and bedding for comfort, and you have a cost-effective shelter that provides protection and warmth for your animals.

Homesteading Clothing and Gear

In addition to materials for projects, thrift stores are also great places to find **workwear** and **protective clothing**. Gardening and livestock care can be tough on clothes, so finding durable, second-hand overalls, boots, and gloves at a low cost can be a practical solution. Often, you'll find that these clothes are already broken in, making them comfortable to wear during long hours of work.

You can often find practical solutions like **old cabinets** or **plastic storage bins** at thrift stores, which can be repurposed to store seeds, tools, or small gardening supplies. Proper storage helps keep your homestead organized and ensures that your tools and supplies are easy to access when needed.

The Value of Thrifting for Sustainability

Beyond the cost savings, shopping second-hand promotes sustainability by keeping materials out of landfills and giving them a second life. It is a great way to reduce waste while creating a functional and unique homestead. As you explore thrift stores and yard sales, keep an open mind—many items can be upcycled with just a bit of creativity.

By embracing these hacks, you can reduce the costs of setting up and maintaining your homestead while contributing to a more sustainable lifestyle.

Creative Ways to Save Energy and Resources

Reducing energy and resource depletion is an essential part of sustainable homesteading. Here are some creative ways to save energy and resources without sacrificing productivity or comfort.

Energy-Efficient Lighting

. . .

One of the simplest ways to cut down on electricity use is by adopting energy-efficient lighting solutions. Replacing traditional incandescent bulbs with *LED bulbs* can significantly reduce your energy consumption, as LEDs use up to 75% less energy and last much longer. Solar-powered lights are also a great option, especially for outdoor spaces like pathways, gardens, and chicken coops. By harnessing the sun's energy, solar lights require no electricity and are perfect for homesteads in off-grid or rural areas.

Solar-Powered Devices

Beyond lighting, consider incorporating solar-powered devices to reduce your reliance on the electrical grid further. Solar-powered water heaters, electric fences for livestock, and small solar panels for charging devices like phones or garden equipment can help you tap into renewable energy. While solar panels for powering your entire homestead can be a larger investment, starting with small-scale solar projects can make a noticeable difference in your energy use. For example, installing a *solar-powered fan* in your barn or greenhouse helps maintain airflow without increasing your electric bill.

Insulation and Weatherproofing

One of the biggest drains on energy in any home is heating and cooling, especially in older buildings. Invest in proper insulation for your home, barn, or animal shelter to reduce energy loss. Insulating your walls, floors, and attics can help maintain a stable indoor temperature, reducing the need for heating in winter and cooling in summer.

For homesteads in colder climates, weatherproofing windows and doors by sealing gaps and adding *draft stoppers* can also save a significant amount of energy. Even simple steps like hanging *thermal curtains* or using *rugs* to insulate cold floors can make a big difference.

Water Conservation Techniques

Water is a vital resource on any homestead, but it's also one that can be wasted if not managed properly. In addition to setting up a *rainwater collection system*, there are several ways to conserve water both indoors and outdoors.

- **Greywater Recycling**: Reusing greywater (wastewater from sinks, showers, and washing machines) can reduce your reliance on fresh water. Greywater can be filtered and repurposed for irrigation, helping to water your garden without increasing your water bill.
- **Efficient Irrigation**: *Drip irrigation systems* are an excellent way to water your garden without wasting water. Unlike traditional sprinklers, which lose a lot of water to evaporation and runoff, drip irrigation delivers water directly to the roots of plants, ensuring that they get the moisture they need with minimal waste.
- **Rain Gardens**: Another effective way to manage water runoff and conserve resources is by creating a *rain garden*. This design captures rainwater runoff from surfaces like rooftops and driveways, preventing soil erosion and flooding. A shallow depression in the landscape filled with native plants allows rainwater to filter into the ground, replenishing groundwater supplies and supporting pollinators. Rain gardens also reduce the strain on stormwater systems and are low-maintenance once established.

Composting and Organic Waste Reduction

Reducing your household waste is another important way to conserve resources. Composting kitchen scraps, yard waste, and livestock manure reduces the amount of waste going to the landfill and provides you with valuable *organic fertilizer* for your garden. In addition to the traditional compost bin, consider using the previously mentioned *vermicomposting* system, which uses worms to break down organic material quickly, creating nutrient-rich compost.

By incorporating these energy-saving and resource-efficient practices, you'll reduce your homestead's overall energy consumption, lower your bills, and contribute to a more sustainable way of living. With the same perspective, in the next chapter, we'll explore the art of food preservation, another key skill for any homesteader looking to maximize their harvest and reduce food waste.

6
THE ART OF FOOD PRESERVATION

As a homesteader, one of the most significant advantages of preserving food is the ability to extend your harvest far beyond the growing season. This ensures you can enjoy homegrown fruits and vegetables year-round and saves money by reducing the need to buy out-of-season produce at higher prices. By storing food you've grown yourself, you can better manage your resources and control the quality of what you eat. Let's introduce you to the essential aspects of food preservation.

Canning: Your Step-by-Step Guide

Canning is one of the most valuable skills a homesteader can learn. It allows you to preserve fruits, vegetables, and even meats for long-term storage, ensuring you make the most of your harvest. For beginners, canning can seem intimidating, but with the right approach and basic knowledge, it's a safe, simple, and gratifying way to extend the life of your food.

. . .

Understanding the Difference: Water Bath Canning vs. Pressure Canning

There are two main types of canning:

- **Water Bath Canning**: This method is used for high-acid foods like fruits, jams, jellies, and pickles. It relies on boiling water to safely process jars at 212°F (100°C). The acidity in the food naturally prevents bacterial growth, making this method ideal for preserving items like tomatoes, peaches, or pickles.
- **Pressure Canning**: For low-acid foods such as vegetables (like green beans, carrots, corn, and potatoes), meats and soups, pressure canning is required to reach higher temperatures (above 240°F or 116°C) safely. This ensures that all harmful bacteria are killed, including *Clostridium botulinum* (which can cause botulism). Using a pressure canner is necessary when preserving things like beans, carrots, or poultry.

Step-by-Step Guide to Water Bath Canning

1. **Prepare Your Equipment**: You'll need a large pot for water bath canning, preferably a designated canner with a rack to hold jars off the bottom. Gather your mason jars, lids, bands, and essential tools like a jar lifter and funnel.
2. **Prepare the Food**: Wash, peel, and chop your fruits or vegetables according to the recipe. Blanch fruits and vegetables that need it to maintain their color, texture, and flavor during the canning process. For jams or jellies, cook the fruit with sugar and pectin until it reaches the desired consistency. For pickles, prepare a vinegar brine.

HOMESTEADING ESSENTIALS

3. **Fill the Jars**: Use a funnel to carefully fill each sterilized jar with hot food, leaving about ½ inch of headspace to allow for expansion during processing. This space ensures the jars seal properly without cracking.
4. **Remove Air Bubbles**: To prevent trapped air from affecting the seal, use a non-metallic spatula or chopstick to stir and remove air bubbles from the jar gently. This step is key to maintaining a long-lasting, safe seal.
5. **Apply the Lids**: Wipe the rims of the jars clean with a damp cloth to ensure a good seal. Place the lids on top, then screw on the bands until they are finger-tight—not too loose, not too tight.
6. **Process the Jars**: Submerge the jars in boiling water, ensuring they are covered by at least 1-2 inches of water. Bring the water to a rolling boil, then process the jars for the time specified in your recipe. Altitude adjustments may be necessary, so check your location's requirements.
7. **Cool and Check the Seals**: Once processing is done, use a jar lifter to carefully remove the jars from the canner. Let them cool on a towel for 12-24 hours. During cooling, you'll hear the characteristic popping sound, which signals that the jars are sealing. After cooling, press the center of each lid—if it doesn't flex, the seal is good.

Step-by-Step Guide to Pressure Canning

For low-acid foods, follow these steps:

1. **Prepare Your Pressure Canner**: Unlike water bath canning, pressure canning uses steam under pressure to reach higher temperatures. Add 2-3 inches of water to your pressure canner, place the jars inside, and lock the lid.

2. **Vent the Steam**: Allow steam to vent for 10 minutes before adding the weight. This ensures the air inside is expelled, and the pressure canner reaches the right internal temperature.
3. **Monitor Pressure**: Once the vent is sealed, increase the heat to bring the canner to the required pressure, as dictated by your recipe. Maintain this pressure for the specified time, adjusting based on your altitude.
4. **Cool the Canner**: After processing, allow the pressure canner to cool and release pressure naturally. Do not attempt to open it immediately, as this could cause jars to break or lose their seals.

In the dedicated section on storage, we'll cover how to store your jars and extend their shelf life safely. By following these basic steps, you'll be able to confidently preserve your fruits and vegetables. In the next section, we'll explore other preservation techniques to help you diversify your food storage strategies.

Freezing, Dehydrating, and Pickling Produce

In addition to canning, freezing, dehydrating, and pickling are some of the most practical and flexible methods of preserving your homestead's harvest. These methods allow you to extend the life of your fresh produce and also offer a variety of ways to enjoy your fruits and vegetables throughout the year. Each method has unique advantages and steps. In this section will guide you through the basics of each preservation technique.

Freezing: Simple and Effective

. . .

Freezing is one of the easiest ways to preserve fruits, vegetables, and even herbs. It's a great choice for beginners because it requires minimal equipment and maintains much of the produce's original flavor and nutritional content.

- **Blanching for Best Results**: Before freezing most vegetables, it's important to blanch them. Blanching involves briefly boiling the vegetables and then immediately transferring them to an ice bath. This process helps deactivate enzymes that can cause color, flavor, and texture loss over time. It also helps preserve the food's vitamins and minerals, making it taste fresher when thawed.
- **Flash Freezing for Convenience**: For fruits like berries or chopped vegetables, using the flash freezing method helps prevent clumping. Spread the produce out on a baking sheet in a single layer and freeze it for a few hours before transferring it to freezer bags. This method keeps individual pieces from sticking together, making it easier to grab only what you need when cooking.

Dehydrating: Long-Term Storage with Minimal Space

Dehydrating is a great way to preserve foods that require little space and can last for long periods without refrigeration. It works by removing moisture from the food, preventing bacteria and mold growth. You can use a food dehydrator, which is a specialized appliance that uses low heat and airflow to dry food evenly. Alternatively, you can dehydrate in an oven at a low temperature (around 140°F or 60°C) or even by using the sun, though sun-drying works best in warm, dry climates.

- **Slicing for Consistent Drying**: When preparing produce for dehydration, uniform thickness is key. If the slices are

too thick, they may not dry evenly, and thicker pieces take longer to dehydrate. Aim for slices that are between 1/8 and 1/4 inch thick to achieve consistent results. Use a mandoline or sharp knife to ensure even slicing.
- **Foods Best Suited for Dehydration**: Many fruits and vegetables dehydrate well, including apples, bananas, tomatoes, carrots, and herbs. Dried fruits make excellent snacks, while dehydrated vegetables can be added to soups, stews, or casseroles and rehydrated during cooking.
- **Dehydrating Herbs**: Drying herbs is one of the easiest preservation methods. Herbs like basil, oregano, and thyme retain much of their flavor when dried and can be used year-round in cooking. For best results, dry herbs in bunches by hanging them in a well-ventilated area or laying them on trays in your dehydrator.

Pickling: Add Flavor and Preserve

Pickling is a preservation method and a way to enhance the flavor of vegetables and even some fruits. It involves soaking produce in a brine made of vinegar, salt, and spices. The acidic environment created by the brine prevents spoilage and allows the pickled goods to last for months.

- **Quick Pickling for Beginners**: Consider starting with quick pickling, also known as refrigerator pickling. This method skips the boiling water bath and simply requires storing the pickled goods in the fridge. While quick pickles don't last as long as traditionally pickled items, they're ready to eat within a day or two and are perfect for adding tangy flavor to salads, sandwiches, or snacks.
- **Customizing Your Brine**: One of the best things about pickling is the ability to tailor your brine to your taste. Add

garlic, dill, mustard seeds, or peppercorns to enhance the flavor of your pickles. You can pickle a variety of vegetables, including cucumbers, carrots, beets, and peppers, and experiment with different spice combinations to create unique flavors.

Each food preservation technique offers distinct advantages depending on the produce and how you want to use it later.

How to Store Your Preserved Foods Safely

Once you've put in the effort to preserve your food, proper storage becomes essential to maintain the quality and safety of your preserved goods. Each preservation method requires specific storage conditions to extend shelf life and prevent spoilage. In this section, we'll explore how to store your preserved foods correctly and what to watch out for to ensure your hard work doesn't go to waste.

Storing Canned Foods

Properly stored canned foods can last for over a year, but they need the right environment to remain safe and fresh.

- **Cool, Dark, and Dry**: Store your canned foods in a cool, dark, and dry location, ideally between 50-70°F (10-21°C). Excess heat can cause food to spoil more quickly, while moisture can cause the lids to rust, potentially compromising the seal.
- **First In, First Out (FIFO)**: Label your jars with the date they were canned, and always use the oldest jars first to avoid having items sit for too long. This FIFO method ensures nothing gets forgotten at the back of the shelf.

- **Checking for Seals**: Before storing, check each jar's seal by pressing the lid—if it doesn't flex, the seal is good. Always inspect jars before consuming, especially if you notice any bulging lids or leaks. These are signs of spoilage and indicate the food should be discarded.
- **Signs of Spoilage**: Watch for cloudy liquids, off smells, or mold inside the jars. If you see any of these signs, the food is not safe to eat.

Storing Frozen Foods

We've covered how freezing is an excellent way to maintain your harvest's fresh taste and nutrients. Here's how to make sure your frozen foods stay fresh for months.

- **Proper Packaging**: Use airtight freezer bags or containers to prevent air exposure, which causes freezer burn. For better results, consider vacuum-sealing to remove all air, keeping your food fresher for longer.
- **Optimal Freezer Temperature**: Keep your freezer set to 0°F (-18°C) or lower. This ensures your food remains frozen solid, stopping the growth of bacteria and maintaining quality.
- **Signs of Freezer Burn**: While freezer burn doesn't make food unsafe, it affects taste and texture. If you see grayish or whitish patches, it means air has gotten to the food. These areas can be trimmed off, but it's best to avoid freezer burn by sealing food properly in the first place.

Storing Dehydrated Foods

. . .

Dehydrated foods are perfect for long-term storage because they take up little space and don't require refrigeration. However, proper storage is key to preventing rehydration and spoilage.

- **Airtight Containers**: Dehydrated foods should be stored in airtight containers like mason jars or vacuum-sealed bags. Keeping out moisture is essential, as it can cause the food to spoil.
- **Cool, Dark Storage**: Store your dehydrated goods in a cool, dark place. Ideally, the temperature should be below 60°F (15°C) to maintain the food's quality for a long time. High temperatures or exposure to light can degrade the food's nutrients and flavor.
- **Vacuum-Sealing for Longer Shelf Life**: For extended storage, consider using a vacuum sealer to remove all air from the packaging. This extra step provides a longer shelf life and helps retain the texture and flavor of dehydrated foods.

Storing Pickled Foods

Pickling not only preserves your vegetables but also infuses them with flavor. Here's how to ensure they stay fresh:

- **Refrigeration for Quick Pickles**: Quick pickles (refrigerator pickles) should be stored in the fridge and consumed within a few months. They aren't shelf-stable, so proper refrigeration is crucial.
- **Shelf Storage for Canned Pickles**: Pickled vegetables that have been processed in a water bath can be stored in a cool, dark place, just like other canned goods. These pickles can last up to a year or more.

- **Signs of Spoilage**: If pickles appear cloudy, smell off, or have developed mold, discard them immediately. Always check that jars remain sealed and lids are not bulging before opening.

By following these storage guidelines, you can ensure that your preserved foods stay safe, fresh, and ready to enjoy for months to come. In the next section, we'll explore creative ways to use your preserved goods in everyday meals, helping you make the most of your homestead's bounty.

Creative Uses for Preserved Goods in Everyday Meals

The real value comes from integrating preserved goods into everyday dishes, turning simple ingredients into nutritious meals. Thinking creatively, you can use your stored produce to enhance from breakfasts and snacks to main courses and desserts. Here are some easy and delicious ways to incorporate your preserved goods into your daily routine.

Canned Foods: Sauces, Soups, and Desserts

Canned fruits and vegetables are incredibly versatile and can be used in numerous ways throughout the year.

- **Tomato Sauce and Soups**: If you've canned tomatoes, you have the base for many meals right at your fingertips. Use your homemade tomato sauce as the foundation for a hearty pasta dish, pizza topping, or chili. You can also add your canned tomatoes to soups or stews to give them a rich flavor and extra nutrients. Combine your tomato base with onions, garlic, and canned beans or frozen vegetables to create a nourishing vegetable soup.

- **Fruit Fillings for Desserts**: Canned fruits, such as peaches, cherries, or apples, can easily be transformed into delicious desserts. Use them to make fruit pies, cobblers, or crisps. For a quick and easy dessert, simply heat your canned fruit and serve it over ice cream, yogurt, or pancakes for a burst of natural sweetness.
- **Quick Salsas and Sauces**: If you've canned vegetables like peppers or corn, consider turning them into a quick salsa or relish. Mix them with fresh or frozen herbs, a dash of vinegar or lime juice, and a pinch of salt for a fresh-tasting topping that's perfect for tacos, grilled meats, or as a dip for chips.

Frozen Produce: Quick Stir-Fries, Smoothies, and Breakfasts

Frozen fruits and vegetables retain much of their original texture and flavor, making them ideal for quick and easy meals.

- **Smoothies and Breakfast Bowls**: Frozen berries, bananas, or peaches are perfect for making nutrient-packed smoothies. Blend them with yogurt or milk for a refreshing breakfast or snack. You can also use frozen fruit as a topping for oatmeal, granola, or yogurt bowls, adding a burst of flavor and vitamins to your morning meal.
- **Stir-Fries and Casseroles**: Frozen vegetables are a great addition to quick stir-fries or casseroles. Simply toss them into a hot pan with some oil, garlic, and soy sauce for a quick side dish or mix them into casseroles for extra color and nutrients. Frozen peas, carrots, and broccoli are particularly well-suited for these types of meals.

Dehydrated Foods: Snacks, Baking, and Soups

Dehydrated fruits, vegetables, and herbs are incredibly versatile and can be used in both savory and sweet dishes.

- **Healthy Snacks**: Dehydrated fruits, such as apple or banana chips, make for a healthy and portable snack. Pair them with nuts or cheese for a simple, nutritious treat on the go. Dehydrated vegetables, like kale chips, can also be seasoned and baked for a crunchy snack.
- **Baking Add-Ins**: Dehydrated fruits can be rehydrated and added to baked goods such as muffins, bread, or cookies. Soaking them in warm water for a few minutes brings back their moisture, and they add natural sweetness and texture to your baking.
- **Herbs for Flavoring**: Dried herbs, such as basil, oregano, or thyme, are perfect for seasoning soups, sauces, and roasted vegetables. They retain their flavor long after drying and can be stored for extended periods without losing their potency.

Pickled Vegetables: Snacks, Sandwiches, and Toppings

Pickled vegetables add a tangy flavor to many dishes and are a great way to elevate a simple meal.

- **Toppings for Sandwiches and Burgers**: Use your homemade pickles, pickled onions, or cucumbers as a flavorful topping for sandwiches and burgers. Their tangy bite adds depth to savory dishes and pairs perfectly with meats and cheeses.

- **Side Dishes and Snacks**: Pickled vegetables, like carrots, beets, and cucumbers, can be enjoyed as a side dish or a snack on their own. They are a refreshing addition to a charcuterie board or picnic spread, offering a balance of sweet and sour flavors.

Just as incorporating preserved goods into your meals maximizes the value of your harvest, efficiently managing your time and resources ensures you get the most out of your efforts. In the next chapter, we'll explore how to keep your homestead running smoothly, even with a busy schedule.

7
HOMESTEADING FOR BUSY PEOPLE

As someone who has navigated the world of homesteading while balancing a hectic schedule, I understand the importance of managing time effectively. This lifestyle requires dedication, but it's also possible to integrate it into even the busiest lives by planning carefully and using time-saving strategies. Whether you're juggling a full-time job, family responsibilities, or other commitments, the key is to start small, stay organized, and focus on efficiency. This chapter will provide you with practical tips and methods to manage your homestead without feeling stressed.

Time-Saving Tips for Weekends and Busy Schedules

One of the biggest challenges of homesteading is finding the time to complete all the necessary tasks. The good news is that there are many ways to streamline your homesteading activities, particularly on weekends or during limited free time.

. . .

Prioritizing Tasks and Staying Flexible

When time is limited, the first step is to **prioritize your most essential tasks**. Begin by identifying what absolutely needs to get done each day or week. Feeding animals, watering plants, and handling urgent maintenance are prime examples of tasks that should take priority. By focusing on must-do items first, you'll ensure the health and productivity of your homestead even when you're short on time.

- **Batch Similar Tasks**: Instead of spreading tasks across multiple days, consider batching similar activities into one focused block of time. For instance, designate one day to process harvested produce or tackle all of your garden's weeding at once. Batching reduces task-switching, making it easier to stay productive.
- **Weekend Work**: If you are balancing full-time work, weekends offer the best chance to get larger tasks done. Divide bigger projects like building a raised bed or deep-cleaning the chicken coop into manageable steps. By breaking them up, you can spread the workload over two or more weekends without feeling overwhelmed.

Effective Use of Planning Tools

It's helpful to incorporate basic planning tools to outline daily and weekly priorities to maximize your time. You don't need a full checklist here—just a simple *task schedule* or calendar where you jot down what needs to be done. This keeps you focused without overcomplicating the process.

- **Set Realistic Goals**: Instead of filling your schedule with too many tasks, set realistic daily goals that allow for

flexibility. Focus on what you can comfortably accomplish in the time you have.
- **Delegate When Possible**: If you have family members or friends who are willing to help, don't hesitate to delegate certain tasks. Shared labor not only speeds up work but also lightens the load for busy weekends.

Streamlining Daily Tasks

By streamlining your daily routine, you can reduce the amount of time spent on repetitive tasks:

- **Automated Systems**: Use tools like automatic watering systems in your garden or automatic feeders for livestock to reduce the number of daily chores. These tools free up valuable time without sacrificing the care of your plants or animals.
- **Short, Efficient Work Bursts**: When it comes to daily tasks, adopt the strategy of short, focused work bursts. Rather than spreading tasks across the day, set aside specific blocks where you can concentrate on completing several tasks simultaneously. This technique is advantageous when you have limited time but want to make the most of it.

By applying these time-saving strategies and proper planning, you can maximize your time on weekends and weekdays while balancing homesteading demands. Later in the chapter, we'll cover daily, weekly, and seasonal checklists to help organize tasks.

Efficient Gardening Methods (No-Dig and Low-Maintenance Approaches)

. . .

As a busy homesteader, adopting efficient gardening methods can make all the difference in maintaining a thriving garden without spending hours each day on upkeep. Two popular approaches for minimizing time and effort are *no-dig gardening* and other low-maintenance techniques. These methods reduce physical labor and improve your soil's health, leading to better long-term results with less ongoing effort.

No-Dig Gardening: Working with Nature

The no-dig gardening method is based on the idea of *preserving the natural structure* of the soil by avoiding tilling or digging. This approach saves time and supports soil health by preserving essential microorganisms, earthworms, and the natural layering of nutrients. Traditional digging can disrupt these beneficial organisms and lead to soil erosion, so skipping this step can promote a more fertile and self-sustaining garden.

- **How to Set It Up**: In no-dig gardening, you simply layer organic materials like compost, manure, and directly on top of the soil. These layers break down over time, adding nutrients to the soil while also suppressing weeds and improving moisture retention. You can start a no-dig garden by laying down cardboard or old newspapers to smother existing weeds, then adding layers of organic matter on top.
- **Less Weeding, Less Watering**: One of the main advantages of no-dig gardening is the significant reduction in weeding and watering. Since the soil is covered with mulch or compost, it retains moisture longer, meaning you won't need to water as frequently. The thick layer of organic matter also acts as a natural weed barrier, cutting down on the need for constant weeding.

No-dig gardening is perfect for those who want to reduce their gardening workload while still producing healthy, nutrient-rich crops. It's a low-maintenance system that requires less intervention over time as the soil structure continues to improve naturally.

Low-Maintenance Gardening Techniques

In addition to no-dig gardening, there are several other low-maintenance gardening techniques that can help save time and energy, making it easier to manage your homestead garden alongside a busy schedule.

- **Mulching**: Applying a thick layer of mulch around your plants is one of the simplest and most effective ways to reduce garden maintenance. It helps retain soil moisture, suppresses weeds, and adds nutrients as it breaks down. Organic mulches, such as straw, grass clippings, or leaves, are ideal because they also improve soil structure over time.
- **Perennial Plants**: Growing perennial crops is another great way to reduce garden maintenance. Unlike annual plants, which need to be replanted each season, perennials come back year after year with minimal care. Herbs like mint, oregano, and thyme, as well as crops like asparagus, rhubarb, and berries, are excellent low-maintenance options that produce for multiple seasons without replanting.
- **Self-Watering Systems**: Installing self-watering systems like drip irrigation or soaker hoses can drastically cut down on the time spent manually watering your garden. These systems deliver water directly to the roots of your plants, reducing evaporation and ensuring consistent hydration. With a self-watering system in place, you can set it on a timer or simply let it run as needed, allowing you to focus on other tasks.

HOMESTEADING ESSENTIALS

These approaches minimize the need for weeding, watering, and soil preparation, freeing up more time to focus on other areas of your homestead. In the next section, we'll also explore quick and efficient livestock care routines designed to fit even the most demanding agendas.

Quick Livestock Care Routines

One of the most time-consuming aspects of homesteading can be the care and maintenance of your livestock. However, with a few simple routines in place, you can streamline the daily care of your animals, ensuring they are healthy and well-tended while fitting these tasks into even the busiest schedule. Whether you're caring for chickens, rabbits, or other small livestock, focusing on *efficient routines* will help you save time without sacrificing the quality of care your animals receive.

Automate Feeding and Watering

One of the best ways to cut down on time spent tending to livestock is by automating feeding and watering systems. Many small livestock, such as chickens and rabbits, can thrive with automatic feeders and self-watering systems.

- Installing **automatic feeders** ensures your animals have consistent access to food without requiring you to measure and distribute it daily. Gravity-fed feeders are a simple and inexpensive option, ensuring food is always available for chickens. For rabbits, pellet dispensers work similarly, providing a steady feed supply throughout the day.
- Using **automatic waterers** or **nipples** attached to a bucket system is another time-saving strategy. These systems keep water clean and available to your animals at all

times. Chickens, for example, can use nipple drinkers, which prevent water spillage and contamination. Similarly, rabbits can use gravity-fed water bottles, ensuring they stay hydrated without needing constant refills.

Streamline Cleaning Routines

Maintaining a clean living environment is crucial for the health of your livestock. Still, cleaning tasks can often feel overwhelming, especially when time is limited. Establishing a routine for quick, *regular upkeep* helps prevent major build-up and reduces the need for more intensive cleaning.

- **Deep Bedding Method**: One highly efficient strategy for chickens and other small livestock is the deep litter method. This involves regularly adding fresh bedding (such as straw or wood shavings) on top of the old bedding, allowing it to compost naturally over time. Not only does this reduce odors and the need for frequent cleaning, but it also creates a warm environment for your animals in colder months.
- **Daily Spot Checks**: Instead of doing a full clean every day, do quick spot checks to remove droppings and refresh any dirty bedding. This can take as little as five minutes and ensures that your animals remain healthy without requiring a full deep clean every day.

Routine Health Checks

A key part of livestock care is monitoring the health of your animals. Instead of waiting until problems arise, incorporate *quick daily health checks* into your routine to catch any issues early. These checks can be done as part of your morning or evening chores and only take a few minutes.

- **Visual Inspection**: As you feed and water your animals, take a moment to observe their behavior and physical condition. Look for signs of illness, such as lethargy, loss of appetite, or any visible injuries or abnormalities. Healthy animals should be active, alert, and have clear eyes and a smooth coat.
- **Handling and Grooming**: For animals like rabbits, spending a few minutes each week handling and grooming them can help you detect health issues early, such as overgrown nails or fur matting. This routine care helps maintain their overall well-being without requiring intensive, time-consuming health checks.

You can manage livestock care efficiently even with a busy lifestyle by incorporating all these steps into your daily schedule.

Staying Organized: Daily, Weekly, and Seasonal Checklists

When you're juggling a busy schedule with homesteading tasks, staying organized is key to ensuring nothing falls through the cracks. One of the most effective ways to manage your homestead efficiently is by creating **checklists** that break down tasks into daily, weekly, and seasonal categories. These lists help you stay on top of essential chores, prioritize tasks, and avoid feeling overwhelmed by the ongoing demands of your homestead. By structuring your work this way, you'll be able to plan, use your time wisely, and maintain a productive homestead without added stress.

Daily Checklists: Consistency Is Key

. . .

Daily tasks are the backbone of running a homestead, ensuring everything from feeding animals to checking crops is handled regularly. Keeping a daily checklist helps you focus on small but essential tasks that keep your homestead running smoothly.

- **Animal Care**: This will likely be the most important part of your daily routine, especially if you have livestock. Feeding, watering, and conducting quick health checks should be at the top of your list. For chickens, this might include collecting eggs and refilling feeders and waterers. For rabbits, it could involve ensuring their hutch is clean and well-stocked with fresh food.
- **Gardening**: Daily gardening tasks vary depending on the season but may include watering plants, harvesting ripe produce, checking for pests, and pruning. If you use an automatic watering system, your garden tasks will be quicker, allowing you to focus on monitoring plant health.
- **Quick Cleanups**: Spend 5-10 minutes tidying up areas like the chicken coop, garden beds, or animal pens. This regular attention helps prevent build-up and reduces the need for deep cleans.

Weekly Checklists: Tackling Bigger Jobs

Some tasks don't need to be done every day but still require consistent attention. A weekly checklist is a great way to tackle slightly larger jobs without them piling up.

- **Cleaning Animal Shelters**: Once a week, schedule a more thorough clean of animal coops, pens, or hutches. Remove old bedding, clean out feeders and water

containers, and check for any signs of damage or pests.
- **Composting**: If you're using compost for your garden, spend time each week turning your compost pile to keep it aerated. This helps speed up the breakdown of materials and prevents the pile from becoming too compact.
- **Garden Maintenance**: Set aside time each week for more extensive garden maintenance, such as weeding, mulching, or preparing beds for planting. Regular weeding is particularly important to prevent overgrowth that could damage your plants.

Seasonal Checklists: Preparing for Changes

In addition to daily and weekly routines, creating seasonal checklists helps you anticipate and manage the shifting demands of your homestead throughout the year. Rather than going into the specifics of each action, this checklist is about organizing your time and ensuring you're ready for the season ahead.

- **Spring**: Plan your planting schedule, prepare your garden beds, and begin sowing seeds indoors or outdoors, depending on your climate. Make sure your tools are cleaned and ready for use, and start your livestock care routine for the warmer months ahead.
- **Summer**: Schedule regular harvesting times and keep track of which crops need attention. Focus on garden maintenance, including watering, weeding, and pest control. Summer can also be the time to make repairs to infrastructure like barns, fences, or coops when the weather is favorable.
- **Fall**: Begin preparing for the colder months. Harvest and preserve the remaining crops, and start planning for food preservation if you haven't already. This is the time to

ensure your livestock shelters are prepared for winter, and your garden beds are cleared for the off-season.
- **Winter**: Winter is often a quieter season on the homestead, making it ideal for planning and indoor projects. Start mapping out next year's garden, organize your seed inventory, and prepare indoor spaces for seed starting. Additionally, ensure your livestock has everything they need to stay warm and healthy through the cold months.

By breaking tasks into these chunks, you can focus on preparing for the upcoming season rather than scrambling to catch up. In the next chapter, we'll focus on how to apply these timing and organization skills to the specific tasks of planting, harvesting, and animal care.

8
SEASONAL HOMESTEADING: WHAT TO DO EACH SEASON

SPRING: Preparing for Growth

Spring marks the start of your homesteading year, bringing renewed energy to the land and new opportunities to build a productive and self-sustaining environment. As the season of growth, it is all about setting strong foundations—both in the garden and for your livestock. With the proper planning and effort, you can ensure a successful and bountiful year ahead.

Setting the Stage

Before you begin planting, preparing your garden space is essential, as this step will determine how well your plants grow throughout the season. Whether you're working with raised beds, containers, or traditional in-ground rows, the groundwork you lay in spring will help your crops thrive.

- **Clearing Debris and Weeding**: The first task is to clean up any leftover debris from winter, such as fallen leaves, dead plants, or weeds. Removing this debris helps prevent pests and diseases that could threaten your crops. Weeding is equally important—this is the time to get ahead of weeds before they start competing with your plants for nutrients and water.
- **Improving Soil Quality**: After cleaning up, focus on enhancing your soil's health. Begin by testing your soil to check its nutrient levels, pH, and organic matter content. Depending on the results, you can amend your soil with compost, manure, or specific fertilizers to ensure it's nutrient-rich. This step is vital for new gardeners and experienced homesteaders alike, as nutrient-rich soil leads to stronger, healthier plants.
- **Preparing Garden Beds**: For those using raised beds or containers, spring is the ideal time to add fresh soil or compost to replenish the nutrients that were depleted during the previous growing season. If you're working with traditional in-ground beds, now is the time to till the soil lightly, break up any clumps, and create rows or planting areas.

Starting Seeds: Jump-Starting Your Growing Season

One of the most exciting parts of spring is starting seeds, a critical step for many crops that need a longer growing season. You can do it indoors or directly outdoors, depending on the plant type and your local climate.

- **Starting Seeds Indoors**: For crops like tomatoes, peppers, and eggplants, which need a head start, begin by sowing seeds indoors 6-8 weeks before the last expected

frost. Use seed trays or small pots filled with seed-starting mix, and place them in a warm, sunny location or under grow lights. Be sure to keep the soil moist but not waterlogged. Once seedlings have developed a few sets of true leaves, they can be hardened off and transplanted outdoors.
- **Direct Seeding Outdoors**: Many hardy vegetables, such as peas, spinach, and carrots, can be directly sown into the garden as soon as the soil is workable. Make sure to read the seed packet instructions to know the right depth and spacing for each crop. If you're in a region with a short growing season, consider using row covers or a cold frame to protect young plants from late frosts and extend the growing season.

Building or Repairing Animal Shelters

Spring is also the perfect time to focus on livestock and their housing needs. Whether you're preparing for new animals or maintaining the ones you already have, ensuring that your animal shelters are ready for the warmer months is crucial for their health and productivity.

- **Inspecting and Repairing Structures**: Start by inspecting existing animal shelters—such as chicken coops, rabbit hutches, or bee hives—for any damage caused by winter weather. Look for signs of wear and tear, like leaks, broken fencing, or drafts. Early spring is the time to repair these issues, ensuring your livestock have safe, comfortable homes.
- **Building New Shelters**: If you're expanding your homestead or introducing new livestock, spring is ideal for building new animal enclosures. When planning your new structures, remember to consider factors like ventilation,

predator protection, and ease of access for feeding and cleaning. Early preparation will ensure your animals are safe and secure throughout the year, whether it's a small coop for chickens or a larger pen for goats.
- **Starting Beekeeping**: If you are interested in beekeeping, spring is the time to establish a hive. Order your bees early in the season and prepare your hive structure, ensuring it's placed in a sunny, sheltered spot on your property. Bees are crucial for pollination, and starting a hive in spring will help support your garden's health while providing honey later in the year.

Properly planning and executing these tasks will make the rest of the year far smoother and more successful.

SUMMER: Nurturing and Managing

Summer is the height of activity on the homestead. With longer days and warmer temperatures, it's the time when your hard work in spring begins to pay off, but it also brings new challenges. This season requires consistent care for your crops, a well-timed harvest, and diligent attention to your livestock's health as they deal with the heat. It's a busy but rewarding time.

Tending to Crops: Nurturing Growth in High Heat

As summer temperatures rise, so does the need for careful crop management. Ensuring your plants receive the proper care during this intense growing season is key to a successful harvest.

- **Watering Wisely**: Summer heat means your crops will need more water than usual, but how and when you water is just as important as how much. The best time to water is in the early morning or late evening, when temperatures are cooler, to prevent evaporation. This also reduces the risk of scorching the plants under the midday sun. Consider installing a drip irrigation system or soaker hoses, which deliver water directly to the roots, minimizing water waste.

- **Mulching for Moisture Retention**: To help your soil retain moisture and keep the root systems of your plants cool, apply a thick layer of mulch around your crops. As previously mentioned in this guide, mulching reduces water evaporation, helps suppress weeds, and keeps soil temperatures consistent. Organic ones like straw, grass clippings, or shredded leaves are ideal, as they break down over time and add nutrients to the soil.
- **Pest Control**: Summer is prime time for pests, so it's essential to monitor your crops for signs of infestation. Common summer pests like aphids, beetles, and caterpillars can wreak havoc on your garden if not managed quickly. Companion planting, using natural deterrents like neem oil, or introducing beneficial insects like ladybugs can help control pest populations without the need for harsh chemicals.
- **Supporting Growth**: Many summer crops, such as tomatoes, peppers, and beans, grow rapidly and may need support. Installing trellises, stakes, or cages early in the season helps plants grow upright and reduces the risk of disease by improving air circulation. Keeping plants off the ground also makes harvesting easier and prevents damage from pests or rot.

Harvesting Early Produce: Enjoying the Fruits of Your Labor

. . .

By mid-summer, many early crops will be ready for harvest. Knowing when and how to do it ensures your produce's best flavor and shelf life.

- **Timing the Harvest**: Harvesting at the right time is crucial for flavor and nutrition. For leafy greens like lettuce and spinach, it's best to harvest them in the morning, when their moisture content is highest. Fruits like tomatoes and peppers should be picked when they reach their full color and firmness but before they become overripe. Early crops like zucchini, cucumbers, and peas often need to be harvested frequently to encourage continuous production.
- **Harvesting Techniques**: Use clean, sharp shears or knives to harvest your crops, and be gentle when handling them to avoid bruising or damaging the plant. For delicate fruits like berries or tomatoes, it's best to cradle them as you pick, reducing the risk of splitting or crushing. After harvesting, cool your produce quickly by placing it in the shade or refrigerating it, especially in hot weather.
- **Succession Planting for Continued Harvests**: As you harvest early crops, you may want to practice succession planting to keep your garden productive. Once an early crop like peas or lettuce is finished, you can replant the space with a second crop, such as beans or late-season greens. This ensures a continuous harvest well into the fall.

Caring for Livestock in the Heat: Keeping Animals Healthy

Summer can be particularly stressful for livestock, as heat can cause health problems if not properly managed. Taking steps to keep your animals cool and hydrated is crucial for their well-being.

- **Providing Shade and Ventilation**: Ensure your animals have access to shade throughout the day. If natural shade

from trees or buildings isn't available, create temporary shelters using tarps or shade cloth. Good ventilation in animal enclosures is also essential—consider installing fans or improving airflow by opening windows or vents to prevent heat buildup.
- **Water, Water, Water**: Livestock need constant access to fresh, clean water, especially during the hot summer months. Check water supplies frequently to make sure they aren't running low or becoming warm. Consider adding extra water sources or automatic waterers to ensure your animals stay hydrated. For animals like chickens, adding electrolytes to their water can help them stay cool and avoid dehydration.
- **Monitoring for Heat Stress**: Keep a close eye on your animals for signs of heat stress, which can include panting, drooling, lethargy, or reduced appetite. If you notice these symptoms, take action immediately—move animals to a cooler area, provide additional water, and, if necessary, cool them down with water misting or fans. For poultry, it's crucial to make sure their coop has plenty of ventilation and isn't overcrowded, as chickens are particularly prone to heat stress.

While the heat can be challenging, remember that summer is also a time of abundant growth and reward on the homestead.

FALL: Harvest and Preservation

Fall is a time of transition on the homestead. As the weather cools and the days shorten, it's time to shift your focus from growth to harvesting and preparing for the colder months ahead. It is often one of the busiest seasons as you work to preserve the fruits of your labor and ensure your homestead is ready for winter. From harvesting crops and canning your bounty to winterizing your animal shelters, fall sets the stage for the quieter months to come.

Harvesting: Bringing in the Bounty

Fall is harvest season, the time when your hard work in the garden comes to fruition. Proper timing and handling ensure you get the best flavor and quality from your crops.

- **When to Harvest**: Knowing when to harvest is key to getting the most out of your produce. Root vegetables like carrots, potatoes, and beets can be harvested once the tops begin to die back, while winter squash and pumpkins should be harvested before the first hard frost. Look for signs of maturity, such as deep color and firm texture, as an indicator that your crops are ready. Fall is also the time to harvest herbs before they go dormant for the winter.
- **Harvesting Techniques**: Use sharp garden shears or knives to harvest crops cleanly, especially for larger vegetables like squash. Root vegetables can be gently dug up with a garden fork to avoid damaging them. Harvest crops in the cool of the morning or evening to preserve their freshness, and avoid harvesting during hot midday hours, which can cause produce to wilt.
- **Extending the Harvest**: For some crops, you can extend the harvest into late fall by using row covers or a cold frame to protect them from early frosts. Vegetables like kale, Brussels sprouts, and spinach thrive in cooler temperatures and can be harvested even after a light frost, which often improves their flavor.

Canning and Storing Food: Preserving Your Harvest

HOMESTEADING ESSENTIALS

. . .

Once you've harvested your crops, the next step is to preserve them so that you can enjoy fresh, homegrown food throughout the winter. Fall is a busy time for canning, freezing, dehydrating, and storing produce.

- **Canning the Harvest**: Canning is one of the most reliable methods for preserving the abundance of your garden. Now is the time to can tomatoes, apples, pears, and root vegetables. Don't forget to follow the detailed steps to canning discussed earlier in the book.
- **Storing Root Vegetables**: Root crops like potatoes, carrots, and onions are perfect for cold storage. After harvesting, cure your vegetables by letting them dry in a cool, well-ventilated space for several days. Once cured, store them in a cool, dark place with good air circulation, such as a root cellar or a cool basement. Avoid storing them near apples or pears, as these fruits emit ethylene gas can cause the vegetables to spoil more quickly.
- **Freezing and Dehydrating**: For crops you don't want to can, consider freezing or dehydrating. Vegetables like beans, peas, and corn freeze well, while fruits like apples and pears can be dried and stored for snacking or baking throughout the winter. Remember to blanch vegetables before freezing to retain their flavor and texture, and store them in airtight bags to prevent freezer burn.

Preparing the Homestead for Winter

With the harvest complete, it's time to shift your attention to winterizing your homestead. Fall is the time to ensure that your garden, tools, and animal shelters are ready for the colder months ahead.

- **Winterizing the Garden**: After harvesting, it's important to clean up the garden to prepare it for the off-season. Remove dead plants, compost what you can, and mulch bare soil to protect it from erosion and nutrient loss during winter. For raised beds, adding a layer of straw or cover crops helps enrich the soil and prevent weeds from taking over before spring.
- **Protecting Livestock**: Ensure your animal shelters are prepared for winter weather. Clean out coops, barns, or hutches thoroughly and add extra bedding to provide warmth. Check for drafts or leaks, repair any damage to the structures, and ensure the shelters are well-ventilated but protected from cold winds.
- **Storing Tools and Equipment**: Finally, fall is the time to care for your tools and equipment. Clean, oil, and sharpen tools before storing them away for winter. Drain fuel from gas-powered tools like mowers or tillers to prevent damage from freezing temperatures. Proper maintenance now ensures your tools are ready for use when spring arrives.

By following these guidelines, you'll enjoy the fruits of your labor throughout the winter and set yourself up for success when the growing season starts again in spring.

WINTER: Maintenance and Planning

Winter is the quiet season on the homestead, but that doesn't mean there's nothing to do. While the colder months are a time of rest and reflection, it's also the perfect opportunity to prepare for the upcoming year. It entails ensuring your livestock is safe and warm, experimenting with indoor gardening projects, and laying the groundwork for the next growing season. By making the most of this time, you'll be ready to hit the ground running when spring arrives.

Winterizing Livestock Shelters: Keeping Animals Safe and Warm

Your animals rely on you to provide them with warmth and protection during the coldest months of the year. Winterizing their shelters is crucial to keeping them healthy and comfortable.

- **Insulating Shelters**: Proper insulation is key to keeping your animals warm without relying too heavily on artificial heating, which can pose fire risks. Add extra bedding—like straw, hay, or wood shavings—inside coops, barns, and hutches to provide warmth. For added insulation, consider hanging old blankets or installing foam boards along the interior walls, taking care to maintain proper ventilation.
- **Preventing Drafts**: Cold drafts can be dangerous for livestock, especially smaller animals like chickens and rabbits. Seal gaps in windows, doors, and walls to prevent cold air from seeping in, but be sure not to seal the shelter too tightly—good ventilation is important for air circulation and reducing moisture buildup, which can cause respiratory issues. Install windbreaks around outdoor shelters to provide additional protection.
- **Heated Waterers**: Access to fresh water is essential, but in freezing temperatures, water can quickly turn to ice. Heated waterers or water heaters for livestock are invaluable tools during the winter months. Chickens, for example, can easily become dehydrated if their water supply freezes. Keep waterers clean and regularly check that they're functioning correctly.
- **Monitoring for Health Issues**: Winter can be tough on livestock, so monitoring your animals regularly for signs of cold stress is essential. Look for behavioral changes such as lethargy, loss of appetite, or huddling in one spot. Provide

extra feed during cold spells, as animals burn more energy to stay warm. For outdoor animals like goats or cows, ensure they have access to a dry, sheltered space to escape snow or freezing rain.

Indoor Gardening: Keeping the Green Thumb Active

While your outdoor garden may be resting under a blanket of snow, winter is the perfect time to start indoor gardening projects. It allows you to grow herbs, leafy greens, and even some vegetables year-round, no matter the weather outside.

- **Starting with Herbs and Greens**: Many herbs, such as basil, parsley, and cilantro, thrive indoors, making them ideal for small-scale winter gardening. Leafy greens like lettuce, spinach, and kale can also be grown indoors with relative ease. Choose a sunny windowsill or set up a grow light system to give your plants the light they need. Make sure to use containers with good drainage and a high-quality potting mix.
- **Hydroponic and Vertical Gardens**: Winter is a great time to experiment with hydroponic systems or vertical gardening indoors. Remember hydroponics allows you to grow plants without soil, using nutrient-rich water instead, while vertical gardening makes the most of limited indoor space. Both methods are efficient and low-maintenance once set up, and they offer the added benefit of year-round fresh produce.
- **Starting Seeds for Spring**: Winter is also the time to start seeds indoors for your spring garden. Some vegetables, such as tomatoes, peppers, and eggplants, require a longer growing season, so starting them indoors 6-8 weeks before the last frost gives them a head start. Invest in a seed-

starting kit or make your own by repurposing containers and using a seed-starting mix. Ensure they get enough warmth and light to grow strong before transplanting them outdoors in the spring.

Planning for Next Year: Laying the Groundwork for Success

Winter provides an opportunity to reflect on the past year's successes and challenges while laying the groundwork for the year ahead.

- **Reviewing the Past Season**: Take time to assess what worked well in your garden and homestead this past year, and what didn't. Did certain crops thrive while others struggled? Were there unexpected challenges, such as pests or weather issues? Use this time to document lessons learned and consider changes to your approach for the upcoming season.
- **Planning the Next Growing Season**: Winter is the perfect time to map out your next garden layout and crop rotation plans. Decide which crops you want to grow and where you'll plant them, taking care to rotate crops to prevent soil depletion and reduce pest and disease problems. Order seeds early to ensure you get the varieties you want, and consider experimenting with new crops or gardening techniques.
- **Organizing and Preparing Tools**: Ensure that your gardening tools are clean, sharpened, and ready for spring. Winter is also a great time to repair or build new infrastructure, like raised beds, compost bins, or animal enclosures. Taking care of these tasks in the off-season means less stress once the busy spring months arrive.

While winter may seem like a quieter time on the homestead, it's actually a season of preparation and planning. This way, you'll ensure you're ready for a productive spring and beyond.

With the seasonal tasks behind you and winter preparations in place, it's time to consider the resilience of your property in the face of unforeseen challenges. In the next chapter, we'll explore how to prepare for emergencies and secure your home and land—ensuring that you're ready to weather the storm no matter what comes your way.

9
EMERGENCY PREPAREDNESS AND HOMESTEAD SECURITY

Stockpiling Essentials for Emergencies

Being prepared for emergencies is a crucial part of homesteading, ensuring that your family and property can weather unexpected events, from natural disasters to power outages. A well-thought-out stockpile provides peace of mind and keeps your household running smoothly during tough times. When building your emergency reserve, focus on essential supplies that will help you stay self-sufficient for an extended period.

Nonperishable food and a reliable **water supply** are at the core. Store enough water for each person—at least *one gallon per day per person* for drinking and sanitation purposes, and aim to have *a minimum two-week supply*. For food, focus on items with long shelf lives, such as canned vegetables, dried beans, pasta, rice, and freeze-dried meals. Rotate your stock regularly to ensure freshness.

Ensure you have a well-stocked **first-aid kit** and **basic medical supplies**, including bandages, antiseptics, pain relievers, and any prescription medications your family might need. Consider adding items like a *thermometer, antibiotic ointment*, and medical gloves for added preparedness.

Power outages are common during emergencies, so having a **backup power source** like a generator or solar charger is essential. We'll delve into this aspect in a dedicated section later in this chapter.

Focusing on these essentials ensures your homestead remains functional and safe during an emergency.

Protecting Your Homestead: Basic Security Measures

Securing your homestead is just as important as storing essentials for emergencies. Taking steps to protect your property safeguards your investments and ensures that your family feels safe and secure. Basic security measures can significantly reduce the risk of theft, vandalism, or intrusion, especially during times of crisis when resources may be limited.

A strong, well-maintained **fence** around your property serves as the first line of defense. Fencing helps keep out unwanted visitors and protects your livestock from predators. Consider installing a **lockable gate** at the main entry point and any additional access areas to secure the perimeter further.

To enhance lighting and visibility, consider installing **motion-sensor lights** around your home, outbuildings, and key areas of your property; it is an effective deterrent against trespassers. Make sure pathways, entrances, and places where valuables are stored, such as barns or sheds, are well-lit. Lights triggered by motion can startle potential intruders and alert you to their presence.

In today's world, technology plays a significant role in home security. Consider installing **security cameras** at strategic points around your homestead, particularly near entry points, gates, and outbuildings. Even simple **alarm systems** or door alarms on outbuildings can provide an extra layer of security and give you peace of mind, especially when you're away from home.

By implementing these basic security measures, you'll create a safer environment for both your family and your homestead.

Power Backup Solutions

Power outages can be one of the most disruptive aspects of any emergency, especially when your homestead relies on electricity for essential systems like heating, water pumps, refrigeration, and communication. Ensuring you have a reliable power backup solution is crucial for maintaining basic functions during a grid failure or extended interruption.

A **portable or standby generator** is one of the most common and effective options. Generators come in a variety of sizes and can power anything from a few essential appliances to your entire home. Portable ones are generally more affordable but require manual setup and fuel management, while standby ones are more expensive but automatically kick in during an outage, offering a seamless power supply. Make sure to keep **extra fuel** on hand, especially if you're in a rural area where fuel deliveries may be limited during a crisis.

A more sustainable and long-term solution is installing a **solar power system** with battery storage. Solar panels allow you to generate your own power, reducing reliance on the grid. Pairing with a battery bank, you can store excess energy for use during power outages. While the initial investment can be higher, solar systems are quiet, eco-friendly, and can provide continuous energy in sunny conditions, making them a great long-term solution for homesteads.

For smaller backup needs, **power inverters** and **battery banks** can be used to store energy and run low-wattage appliances or recharge devices. While they won't power large systems, they are ideal for keeping phones, small lights, or essential communication devices running during shorter outages.

By selecting the right power backup solution for your needs, you'll ensure your homestead remains functional.

. . .

Preparing for Natural Disasters and Extreme Weather

Natural catastrophes and extreme weather events are unpredictable, but with proper planning, you can significantly reduce their impact on your homestead. Each type of disaster requires a unique approach, from floods and storms to droughts and wildfires. Preparing in advance will help protect your property, livestock, and family, ensuring you can weather any storm.

The first step in preparedness is to **understand the specific risks** your homestead faces based on your region. If you live in an area prone to *flooding*, evaluate the elevation of your home and outbuildings. If you're in a region at risk for *wildfires*, make sure to create defensible space around your property by clearing vegetation and using fire-resistant materials.

Develop a **comprehensive emergency plan** that includes evacuation routes, communication strategies, and key actions to take in the event of an emergency. Make sure all family members know the plan and practice drills regularly. For natural disasters like hurricanes or tornadoes, identify safe areas within your home, such as basements or storm shelters, where you can take cover if needed.

Safeguard your livestock by ensuring they have secure shelters and enough food and water to last through the emergency. Reinforce barns, coops, and sheds against high winds, heavy rains, or snow. Consider installing *flood barriers* or *storm shutters* to safeguard your home and outbuildings from extreme weather conditions.

In addition to food and water, **stockpile supplies** like *sandbags, tarps, ropes,* and *emergency repair kits* to deal with immediate damage. These items can help prevent further destruction and allow you to make temporary repairs while waiting for the storm to pass.

By assessing your risks and preparing for worst-case scenarios, you'll ensure that your homestead is as resilient as possible, no matter what nature throws your way.

Now, it's time to build new skills and enhance your self-sufficiency. In the next chapter, we'll dive into how you can develop advanced techniques, explore new crafts, and increase your overall productivity, helping you take your path to the next level.

10

GROWING YOUR HOMESTEADING SKILLS

As you continue your homesteading journey, thinking about how you can grow and expand your skills and operations is natural. The following sections offer an overview of potential developments and opportunities you may explore over time. While these topics open the door to new techniques and ideas, they are just introductions to broader subjects that deserve deeper exploration. The goal here is to give you a glimpse into the possibilities, with the understanding that these abilities can be honed further through dedicated learning and hands-on experience.

Expanding Your Garden and Livestock: Intermediate Skills

Once you've mastered the basics, the next step is expanding skills that will allow you to maximize productivity, improve efficiency, and diversify your homestead, making it more self-sufficient and sustainable.

. . .

HOMESTEADING ESSENTIALS

Expanding Your Garden: Crop Rotation and Companion Planting

As your garden grows, it's essential to practice *crop rotation* and *companion planting* to maintain soil health and prevent pest issues.

- **Crop Rotation**: This technique involves rotating different plant families through your garden beds each season, preventing nutrient depletion and reducing the risk of soil-borne diseases. For example, after growing nitrogen-fixing legumes like beans and peas, plant nitrogen-demanding crops like tomatoes or corn in the same bed the following season. This method helps maintain soil fertility and promotes healthy plant growth.
- **Companion Planting**: This strategy involves growing plants together that benefit each other. For example, planting basil next to tomatoes can enhance the flavor of tomatoes and deter pests like aphids. Similarly, marigolds are excellent companions for most vegetables because they repel harmful insects while attracting beneficial pollinators. Companion planting helps create a more balanced ecosystem in your garden, boosting plant health and yield.

Expanding Livestock: Intermediate Care and Diversification

If you've already mastered the basics of raising small livestock like chickens or rabbits, consider diversifying your animal operations with intermediate livestock or scaling up your current setup.

Consider adding **goats** or **sheep** to your homestead for intermediate livestock care if you have proper space. Both animals are relatively easy to care for and provide multiple benefits—goats can supply milk and meat, while sheep offer wool and meat. These animals require secure fencing and basic shelter, but once you have these systems in place, they're excellent additions to any homestead.

If you've been raising chickens, try expanding into *advanced poultry management* techniques like breeding your own flock or raising other types of poultry, such as **ducks** or **turkeys**. This increases your production and allows you to work toward a more self-sustaining flock by hatching chicks on-site.

Expanding your garden and livestock with intermediate skills is also an excellent time to delve into new crafts that support a self-sufficient lifestyle.

Learning New Crafts: Soapmaking, Herbal Remedies, and Sustainable Cleaning Products

As you grow in this adventure, learning new handmade abilities like soapmaking, creating herbal remedies, and making sustainable cleaning products enhances your self-sufficiency and allows you to reduce reliance on store-bought items and eliminate harmful chemicals from your home. These intermediate skills bring additional value to your experience and provide natural alternatives to everyday products.

Soapmaking: A Practical and Rewarding Craft

Making your own soap is both practical and rewarding, offering a way to create custom, chemical-free products tailored to your needs. **Cold-process soapmaking** is one of the most popular methods, requiring just a few basic ingredients. Here's a simple guide to start crafting your first soap.

Materials Needed:

- **Oils** (e.g., olive oil, coconut oil, or lard)
- **Lye** (sodium hydroxide)

- **Water**
- **Essential oils** (for fragrance, optional)
- **Herbs or exfoliants** (optional)
- **Protective gear** (gloves and goggles)

How to Proceed:

1. **Prepare Your Workspace**: Soapmaking involves lye, (sodium hydroxide), which initiates the saponification process and can be dangerous if improperly handled. Wear protective gear (gloves and goggles), and make sure your workspace is well-ventilated.
2. **Measure and Mix Oils**: Weigh your oils according to your chosen recipe. Common beginner soap recipes often include a mix of olive and coconut oil. Melt the oils if necessary and set aside.
3. **Prepare Lye Solution**: In a separate container, carefully add lye to water (never water to lye) and stir until dissolved. Let the lye solution cool to around 100–120°F.
4. **Combine Oils and Lye**: Once both the lye solution and oils have cooled to similar temperatures (100–120°F), slowly pour the lye into the oils. Stir until it reaches a **trace**, meaning the mixture has thickened slightly and leaves a trail when drizzled on itself.
5. **Add Extras**: One of the joys of soapmaking is the ability to *customize* it to your preferences. You can stir in exfoliants like *oatmeal* or *coffee grounds* for texture or use skin-soothing herbs like *lavender* or *chamomile* to create gentle, nourishing soap bars.
6. **Mold and Cure**: Pour the mixture into molds and cover with a towel. Let the soap sit for 24–48 hours until solid. Once unmolded, allow the soap to cure for 4–6 weeks before using.

Herbal Remedies: Using Nature for Health

Crafting your own herbal remedies allows you to harness the power of nature to treat minor ailments and promote wellness. Start with **common herbal remedies** using herbs you can grow in your garden. For example, *peppermint tea* helps soothe digestive issues, while *calendula salves* can be applied to cuts and burns to promote healing. You can also experiment with tinctures, teas, and salves using herbs like *echinacea, elderberry,* and *ginger* to support immune health. Here's how to make a few basic remedies:

Materials Needed:

- **Dried or fresh herbs** (e.g., peppermint, calendula, echinacea, ginger)
- **Carrier oils** (for salves, e.g., olive or almond oil)
- **Alcohol** (for tinctures)
- **Glass jars** and **dropper bottles**

How to Proceed:

- *Tincture (e.g., Echinacea or Elderberry):*

Ingredients: Fresh or dried herbs, vodka or brandy.
Steps:

1. Fill a jar halfway with herbs.
2. Cover with alcohol, leaving room at the top.
3. Seal and shake daily for 4–6 weeks.
4. Strain the tincture and store it in a dark dropper bottle.

- *Herbal Tea (e.g., Peppermint):*

Ingredients: Dried peppermint leaves.
Steps:

1. Boil water.
2. Add 1–2 teaspoons of dried leaves to a cup.
3. Pour hot water over the herbs, cover, and steep for 10 minutes.
4. Strain and enjoy.

- ***Salve (e.g., Calendula):***

Ingredients: Dried calendula flowers, olive oil, beeswax.
Steps:

1. Infuse olive oil with calendula by placing flowers in a jar, covering them with oil, and leaving in a sunny window for 4 weeks (or gently heat for 2–3 hours).
2. Strain the oil.
3. Melt beeswax and combine with infused oil in a 1:4 ratio.
4. Pour into small tins or jars and let cool.

Sustainable Cleaning Products: Reducing Chemicals at Home

Making your own sustainable cleaning products helps reduce the number of chemicals in your home while saving money. Simple ingredients like *baking soda, vinegar,* and *essential oils* can be used to create effective and eco-friendly cleaners. Here are a few simple recipes to get started:

Materials Needed:

- **White vinegar**
- **Baking soda**
- **Essential oils** (optional, e.g., lemon, tea tree)
- **Castile soap**
- **Spray bottles and jars**

How to Proceed:

- **All-Purpose Cleaner:**

Ingredients: Equal parts water and vinegar, a few drops of essential oil.
Steps:

1. Mix vinegar and water in a spray bottle.
2. Add a few drops of essential oil for fragrance.
3. Use this solution to clean countertops, windows, and floors.
4. **Natural Scrub:**

Ingredients: Baking soda, castile soap.
Steps:

1. Combine baking soda with enough castile soap to make a paste.
2. Use this to scrub sinks, stoves, and other tough surfaces.
3. **Glass Cleaner:**

Ingredients: Water, vinegar.
Steps:

1. Mix equal parts water and vinegar in a spray bottle.
2. Spray on glass surfaces and wipe with a clean cloth.

As you become more skilled, you can also explore how to turn these new abilities—and others—into potential income streams.

How to Start Generating Income from Your Homestead

As your abilities grow, so do the opportunities to generate income from your land and talents. Many homesteaders successfully turn their produce, crafts, and knowledge into income streams that support their self-sufficient lifestyle and provide additional financial security. Whether you're growing vegetables, raising livestock, or crafting sustainable products, there are several ways to monetize your efforts.

Selling Produce and Animal Products

One of the most straightforward ways to generate income from your land is by selling the excess produce you grow or the products from your livestock. **Farmers markets** are a popular venue for selling fresh vegetables, fruits, eggs, honey, milk, and meat directly to consumers. You can also explore **community-supported agriculture (CSA)** models, where customers pre-purchase a share of your harvest and receive weekly or bi-weekly deliveries of fresh produce.

If you're raising chickens, ducks, or goats, selling **eggs** or **goat milk** can provide a steady income. Eggs, in particular, are always in demand, and if you have a surplus, you can sell them to neighbors, at markets, or even to local stores.

Turning your raw goods into **value-added products** is another way to increase your income. For example, instead of just selling fresh tomatoes, you could produce *canned tomato sauces* or *jams*. Similarly, if you're raising bees for honey, you could create *beeswax candles* or *lip balms* to sell in addition to honey.

Selling Crafts and Homemade Goods

Your new skills, such as soapmaking, herbal remedies, and sustainable cleaning products, can also be monetized. These handmade items often appeal to people looking for natural, eco-friendly alternatives to commercial products. You can sell them at **local craft fairs**, online through platforms like **Etsy**, or even set up your own **homestead store** if you have consistent traffic to your property.

Herbal Remedies like *herbal tinctures, salves,* and *teas* can be sold both online and in local markets. As people seek out natural wellness solutions, products made from herbs you grow can become a valuable niche.

Furthermore, sharing your knowledge by hosting **workshops** or **online courses** on topics like **gardening**, **livestock care**, or **soap-making** can be another income stream. Many people are eager to learn homesteading skills, and providing educational opportunities can be both fulfilling and profitable.

With your homestead now working as a potential source of income, it's time to continue growing your skills and exploring new ideas.

Experimenting with New Homesteading Techniques

As you become more comfortable with the foundational practices, it's natural to start exploring new techniques to make your homestead more efficient, sustainable, and productive. Experimentation is key to evolving as a homesteader, helping you fine-tune your methods and discover innovative ways to improve your land and lifestyle.

Permaculture Principles

One of the most exciting approaches to explore is permaculture, a holistic system of design that mimics the natural ecosystems. It focuses on using resources efficiently while creating a self-sustaining environment. For example, instead of traditional gardening, you might try **forest gardening**, which involves planting trees, shrubs, herbs, and vegetables in layers, much like a natural forest. This method reduces the need for watering, fertilizing, and weeding while also providing habitat for wildlife.

Consider experimenting with water management techniques such as **swales** (shallow trenches designed to capture water) or **keyline**

design, which maximizes water distribution across the land. These methods can improve water retention, reduce erosion, and make your land more resilient to drought.

Another approach is **hugelkultur**, a raised garden bed technique that uses logs and other organic materials buried beneath soil to create a long-term, self-composting bed. Hugelkultur beds retain moisture and nutrients better than traditional raised beds, making them ideal for areas with poor soil or water shortages.

Expanding into New Livestock Practices

If you've already mastered basic livestock care, trying new techniques can increase your farm's productivity. Consider introducing **rotational grazing**, a practice where livestock are moved between pastures to allow the land to recover, improve soil fertility, and reduce overgrazing. This technique promotes healthier pastures, leading to better livestock yields and a more sustainable land management system.

For a more integrated approach, explore **silvopasture**, a system where trees and grazing animals coexist. This method combines forestry with grazing, offering shade for animals, improving biodiversity, and boosting long-term land sustainability.

With new techniques to explore, your journey toward greater sustainability continues to evolve. In the next chapter, we'll reflect on how community can play a crucial role in your homesteading experience, offering support, shared knowledge, and collaboration as you grow and learn on your path to self-sufficiency.

11
EMBRACING YOUR ADVENTURE AHEAD

The Power of Connection

While homesteading encourages independence, it's important to remember that you're not alone on this path. Connecting with others —whether in person or online—brings invaluable support, advice, and inspiration. One of the most accessible ways to start is by joining **online communities**. Look for homesteading forums or social media groups where you can interact with others who share your goals. These platforms are invaluable for asking questions, sharing successes and challenges, and learning from others' experiences. You might also want to subscribe to homesteading newsletters or listen to podcasts to deepen your knowledge. Engaging with these online spaces can give you immediate access to advice and resources.

Locally, attending homesteading **workshops**, **farmer's markets**, or **skill-sharing events** can offer hands-on learning opportunities. Participating in these events allows you to build practical skills while also meeting like-minded individuals in your community. Volunteering at a community garden or participating in a local food swap event also provides a chance to connect while exchanging resources.

Additionally, **bartering** is a powerful tool for homesteaders. Establishing a local network for trading goods and services can help you access what you need without money changing hands. For example, you might trade homegrown produce for help with a construction project or share tools to reduce costs. Starting a tool-sharing program in your neighborhood is a practical way to build relationships and lower expenses.

Mentorship is another valuable aspect of connection. Learning directly from an experienced homesteader can save you time and prevent costly mistakes. If you're more experienced, offering to mentor others through workshops or informal meetups can reinforce your own skills while fostering community bonds. Organizing a monthly homesteading roundtable in your area is a great way to share tips and challenges with others, sparking new ideas and support systems.

If there's no existing network in your area, you can take the initiative to start one. Creating a local or online homesteading group provides a dedicated space for sharing resources, advice, and motivation. You can also explore cooperative projects, like building a communal greenhouse or starting a shared composting system. Collaborating on larger tasks with others helps lighten the load while fostering stronger community ties.

These connections, both online and locally, will not only enrich your homesteading experience but also empower you to thrive through shared knowledge and resources.

You're on the Right Path—Keep Going!

As you reach the end of this guide, it's important to remember that homesteading is more than a set of skills or a collection of tasks; it's a way of life that invites you to grow, adapt, and embrace the process of becoming more self-sufficient. You've come a long way, but this is just the beginning. Every season will bring new lessons, challenges, and opportunities for growth. It's not about getting everything perfect from the start—it's about being open to learning and evolving along the way.

There will be successes that fill you with pride and setbacks that teach you resilience. Each challenge becomes an opportunity to learn, and each victory brings you closer to the lifestyle you're building. Self-sufficiency is a process that unfolds day by day, season by season. Embrace that, and you'll always find yourself moving forward, no matter what.

Remember, every step forward, no matter how small, is progress. The garden you plant, the animals you raise, the skills you master—all of these contribute to a more fulfilling, resilient life. On the tough days, remind yourself why you started. You have everything it takes to continue building the homestead and life you envisioned. Keep moving forward with confidence, knowing that each day brings you closer to your goals. Keep exploring, stay curious, and challenge yourself to try something new each year. Your homestead will reflect your growth, evolving along with you.

You've got this!

Thank you for taking the time to read this book. I truly hope it has inspired and empowered you to take the next steps on your homesteading journey. It's been a privilege to share this knowledge with you, and I'm excited to think about how you'll apply these skills to start building the life you envisioned.

If you found this book helpful, I'd be incredibly grateful if you could take a moment to leave a review. Your feedback helps me continue sharing this knowledge with others and provides valuable insight for future readers who are ready to begin their adventure.

Thank you again for your support, and I wish you all the best on your homesteading path!

Landon Woodland

www.ingramcontent.com/pod-product-compliance
Lightning Source LLC
Chambersburg PA
CBHW052324220526
45472CB00001B/265